Environmental Barrier Coatings

Environmental Barrier Coatings

Special Issue Editor

Kang N. Lee

MDPI • Basel • Beijing • Wuhan • Barcelona • Belgrade • Manchester • Tokyo • Cluj • Tianjin

Special Issue Editor
Kang N. Lee
NASA Glenn Research Center
USA

Editorial Office
MDPI
St. Alban-Anlage 66
4052 Basel, Switzerland

This is a reprint of articles from the Special Issue published online in the open access journal *Coatings* (ISSN 2079-6412) (available at: https://www.mdpi.com/journal/coatings/special_issues/Environmental_Barrier_Coat?authAll=true).

For citation purposes, cite each article independently as indicated on the article page online and as indicated below:

LastName, A.A.; LastName, B.B.; LastName, C.C. Article Title. *Journal Name* **Year**, *Article Number*, Page Range.

ISBN 978-3-03936-517-3 (Hbk)
ISBN 978-3-03936-518-0 (PDF)

Contents

About the Special Issue Editor

Kang N. Lee is a Senior Research Scientist at the NASA Glenn Research Center, Cleveland, OH, USA. He received his Ph.D. from the University of Minnesota, Minneapolis, MN, USA, in 1987. After postdoctoral work at the University of Pennsylvania, Philadelphia, PA, USA, he spent the first 15 years of his career at the NASA Glenn Research Center before moving to Rolls-Royce in 2005. After 11 years at Rolls-Royce, he returned to NASA in 2016. During his first tenure at NASA, he pioneered EBC research while leading the NASA EBC programs that developed the first- and second-generation EBCs, which became the foundations for the current state-of-the-art EBCs. As an Associate Fellow at Rolls-Royce, he provided global leadership for strategy and technology development of high-temperature coatings across the Rolls-Royce product line. Since returning to NASA in 2016, his research has focused on cutting-edge coatings R&D, including next-generation EBCs. He has authored 35 US patents, 20 US patents pending, 80 papers in archival journals, and 4 invited book chapters. He is the recipient of the 2017 Toledo Glass and Ceramic Award, 2010 Rolls-Royce Sir Henry Royce Award, 2003 NASA Public Service Medal, and 2001 R&D 100 Award.

Editorial

Special Issue: Environmental Barrier Coatings

Kang N. Lee

NASA Glenn Research Center, Cleveland, OH 44135, USA; ken.k.lee@nasa.gov

Received: 12 May 2020; Accepted: 22 May 2020; Published: 27 May 2020

Abstract: The global increase in air travel will require commercial vehicles to be more efficient than ever before. Advanced turbine hot section materials are a key technology required to keep fuel consumption and emission to a minimum. Ceramic matrix composites (CMCs) are the most promising material to revolutionize turbine hot section materials because of their excellent high-temperature properties. Rapid surface recession due to volatilization by water vapor is the Achilles heel of CMCs. Environmental barrier coatings (EBCs), which protect CMCs from water vapor, is an enabling technology for CMCs. The first CMC component entered into service in 2016 in a commercial engine, and more CMC components are scheduled to follow within the next few years. One of the most difficult challenges to CMC components is EBC durability because failure of EBC leads to a rapid reduction in CMC component life. Novel EBC chemistries, creative EBC designs, and robust processes are required to meet EBC durability challenges. Engine-relevant testing, characterization, and lifting methods need to be developed to improve EBC reliability. The aim of this Special Issue is to present recent advances in EBC technology to address current EBC challenges.

Keywords: EBC; CMC; oxidation; volatility; CMAS; thermomechanical; modeling

1. Introduction

Ceramic matrix composites (CMCs) are considered a game changer for gas turbine hot section materials because of their excellent high-temperature mechanical properties, oxidation resistance, and a light weight—one-third of nickel-based superalloys [1]. The introduction of CMCs enables a fuel burn reduction up to two percent—few other technology in today's pipeline have this much capability for fuel burn reduction [2]. Additionally, the light weight enables over 50 percent reduction in the turbine component weight [2].

SiC/SiC CMCs derive their oxidation resistance from dense slow-growing silica (SiO_2) scale [3]. In combustion environments, however, the protective silica scale volatilizes by reacting with water vapor (H_2O), leading to rapid CMC surface recession [4–9]. Alumina (Al_2O_3) in oxide/oxide CMCs, although not as severe as silica, also suffers surface recession due to water vapor-induced volatilization [10]. External barrier coatings, known as environmental barrier coatings (EBCs), have been developed to protect CMCs from surface recession [11,12]. Over two decades of intense research and development led to the first EBC-coated CMC component-a high-pressure turbine CMC shroud-to enter service in the LEAP engine by CFM International (Cincinnati, Ohio) [2]. More CMC components, including combustor liners and high-pressure turbine vanes, are scheduled to follow in the near future [2].

One of the most difficult challenges to CMC components is EBC durability because failure of EBC leads to a rapid reduction in CMC component life. Key contributors to EBC failure include recession, oxidation, degradation by calcium-aluminum-magnesium silicates (CMAS) deposits, thermal and thermo-mechanical strains, particle erosion, and foreign object damage (FOD) [12,13]. Novel EBC chemistries, creative EBC designs, and robust processes as well as engine-relevant testing, characterization, and lifting methods are required to meet EBC durability and reliability challenges.

2. This Special Issue

The nine research articles in this special issue present recent advances in EBC technology in the following areas: recession [14]; EBC chemistry and processes [15,16]; oxidation and recession [17]; oxygen permeability and EBC design [18]; CMAS thermal and mechanical properties [19]; CMAS infiltration modeling [20]; mechanical behavior in thermal shock [21]; and thermomechanical reliability: modeling and validation [22].

Smialek et al. [14] investigated the volatility of transient TiO_2 and steady-state Al_2O_3 scales formed on Ti_2AlC MAX phase ceramic in 1300 °C high velocity (Mach 0.3, 100 m/s) and high pressure (6 atm, 25 m/s) burner rig tests (BRT). Unlike metals, the ceramic was stable at 1300 °C. Unlike SiC and Si_3N_4, neither burner test produced a weight loss, unless pre-oxidized. Volatility of TiO_2 was indicated by removal of Ti-rich oxide grains after BRT, while Al_2O_3 loss, not as pronounced as TiO_2, was indicated by grain boundary etching and porosity. Al_2O_3 net weight loss was observed only on pre-oxidized samples, whose linear loss rate was about one-fifth of SiO_2. 7YSZ TBC on Ti_2AlC survived for 500 h in the Mach 0.3 burner test with no indication of volatility or spalling.

Vaßen et al. [15] investigated various oxides as an EBC for both oxide/oxide and SiC/SiC CMCs. The goal was to obtain dense and crystalline coatings. APS (atmospheric plasma spraying) Y_2O_3, $Gd_2Zr_2O_7$, $Y_3Al_5O_{12}$, and $Yb_2Si_2O_7$ were investigated for oxide/oxide CMCs, while Y_2SiO_5 and $Yb_2Si_2O_7$ were investigated for SiC/SiC CMCs. Y_2O_3 and $Gd_2Zr_2O_7$ showed promising results for oxide/oxide CMCs: Y_2O_3 coating showed excellent adhesion due to the formation of chemical bond, while $Gd_2Zr_2O_7$ coating required laser structuring of the CMC surface for good adhesion. APS Y_2SiO_5 required a heat treatment to obtain a desirable microstructure for SiC/SiC CMCs. $Yb_2Si_2O_7$ on SiC/SiC CMCs was investigated using various thermal spray processes (APS, high velocity oxygen fuel (HVOF), very low pressure plasma spraying (VLPPS), and suspension plasma spraying (SPS)). APS coating was dense, but with a high degree of amorphous phase content. HVOF gave higher crystalline phase content than APS, but with some porosity. VLPPS coating gave the highest crystallinity and density. SPS gave a high degree of crystallinity, however segmental cracks could not be avoided. The study demonstrated the feasibility of various thermal spray methods besides APS for EBC processing.

Gatzen et al. [16] investigated $YAlO_3$ as an EBC for Al_2O_3/Al_2O_3 CMCs using APS and VLPPS processes. APS resulted in a poor quality coating not suitable as an EBC. VLPPS, on the other hand, resulted in a coating with high crystallinity, high purity, and strong adhesion. The strong adhesion was attributed to the formation of chemical bonding due to the formation of $Y_3Al_5O_{12}$ interfacial phase. The coating exhibited excellent thermal cyclic durability and CMAS resistance in laboratory testing, demonstrating its promise as an EBC for oxide/oxide CMCs.

Klemm et al. [17] investigated the oxidation and corrosion (water vapor-induced SiO_2 volatilization) of two plasma-sprayed EBC systems (Al_2O_3/YAG and $Si/Yb_2Si_2O_7+SiC/Yb_2SiO_5$) on SiC/SiC CMCs. Two EBC failure mechanisms were observed: spallation due to TGO-induced stresses and corrosion-induced gap at the TGO/EBC interface. Al_2O_3/YAG EBC, after burner rig test, suffered corrosion at the TGO/EBC interface and inside the Al_2O_3 bond coat. In $Si/Yb_2Si_2O_7+SiC/Yb_2SiO_5$, TGO-induced failure was delayed because SiC particles in the intermediate layer oxidized first, forming a shell surrounding SiC, and thereby served as a getter to reduce the permeation of oxidants (O_2, H_2O) through EBC. In the burner rig test, small pores developed in the intermediate layer due to SiO_2 corrosion. The beneficial gettering function of SiC particles is temporary, disappearing when the SiC particles are fully consumed.

Kitaoka et al. [18] proposed an EBC design based on oxygen permeability calculations to improve oxygen shielding capability and phase stability of mullite coating in Si/mullite EBC for SiC/SiC CMCs. Grain boundary (GB) diffusion of O ions in mullite, from high PO_2 mullite surface to low PO_2 mullite/Si interface, and simultaneous GB diffusion of Al ions in the opposite direction were calculated. Calculations projected: (i) mullite would decompose at the mullite/Si interface due to the outward Al diffusion and (ii) replacing the Si bond coat with the β′-SiAlON bond coat would increase oxygen shielding capability of mullite—this is because the higher equilibrium PO_2 for the oxidation of β'-SiAlON to form SiO_2 compared to the oxidation of Si to form SiO_2 suppresses the inward O ion diffusion. Experiments confirmed the mullite decomposition at the mullite/Si interface and the improvement of mullite stability when the Si bond coat was replaced by β'-SiAlON. The latter is because β'-SiAlON serves as a source for Al.

Webster et al. [19] investigated the intrinsic thermal and mechanical properties of a volcanic ash glass obtained from the Eyjafjallajökull eruption of 2010. The properties studied include crystallization, melting temperature, CTE (coefficient of thermal expansion), modulus, hardness, and viscosity. The glass had a low propensity to crystallize in bulk form compared to synthetic sand glasses. Compared to sand glasses, the melting temperature and viscosity were higher, while the CTE and modulus were lower. The higher viscosity was attributed partly to a higher SiO_2 and lower CaO content in the volcanic ash. Hardness was similar, but the indentation fracture toughness was about twice that of sand glasses, which was attributed to the presence of Fe_3O_4 crystallites in the volcanic ash. Understanding the variation of CMAS properties with composition can provide insight into CMAS/coating interactions, which can be highly valuable for the development of CMAS-resistant coatings.

Kabir et al. [20] developed a numerical approach for studying the infiltration kinetics of molten CMAS in EB-PVD TBCs. Detailed analyses of the infiltration kinetics were performed, and the results were verified by experimental infiltration depths of the feathery and coarse microstructures. The study identified key morphological features that are important for infiltration. The rate of longitudinal and lateral infiltration could be minimized by reducing the gap between columns and/or increasing the length of the feather arms. Long feather arms having a lower lateral inclination decreased the infiltration rate and, therefore, reduced the infiltration depth. A key takeaway is that the infiltration kinetics can be altered by careful tailoring of coating microstructure such as feather arm lengths and inter-columnar gaps.

Seo et al. [21] investigated crack formation, crack stability, and crack healing of Si/mullite, Si/(mullite + Yb_2SiO_5), and Si/Y_2SiO_5 EBCs on SiC/SiC CMCs under thermal shock cycling between room temperature and 1350 °C in air. Mechanical behavior was determined before and after the test via indentation tests. EBCs developed mud cracks on the surface and unidirectional vertical cracks in the top coat after several thermal shock cycles. Mullite-based EBCs exhibited crack healing phenomena after 4000-5000 cycles, where cracks were covered and partially filled with a new phase. Post-test XRD showed only crystalline Al_2O_3 in addition to the EBC component phases, indicating that the new phase is likely amorphous. The nature of the new phase responsible for cracking healing was not fully understood yet. Si/Y_2SiO_5 EBC delaminated after 4000 cycles with no crack healing phenomena. After 5000 cycles, mullite-based EBCs showed no significant changes in modulus and hardness due to cracking healing, whereas Y_2SiO_5 EBC showed a change from elastic to plastic behavior due to delamination.

Kawai et al. [22] investigated the thermomechanical stability of an as-processed EBC by comparing the ERR (energy release rate) for interfacial cracks calculated by FEM analysis and the interfacial fracture toughness determined experimentally. The EBC studied was (SiC/SiAlON/mullite/Yb disilicate-Yb monosilicate graded layer/Yb monosilicates with porous segmented structure). FEM showed thinner SiAlON/thicker mullite combinations lower the initial ERR. Fracture in the interface fracture test occurred at the SiC/SiAlON and SiAlON/mullite interfaces, indicating these interfaces are the weak link. ERR by FEM was sufficiently higher than the experimentally determined interfacial fracture toughness, indicating that as-processed EBC possesses sufficient thermomechanical reliability. As-processed EBC with 5–25 μm SiAlON and mullite showed no interfacial delamination, confirming the thermomechanical reliability. This study considered only as-processed stresses. Further work is needed to assess the thermomechanical reliability in service environments by incorporating oxidation, CMAS, thermal cycling in temperature gradient, and mechanical stresses.

Funding: This research received no external funding.

Conflicts of Interest: The authors declare no conflict of interest.

References

1. Anson, D.; Richerson, D.W. *Progress in Ceramic Gas Turbine Development Vol. 2*; van Roode, M., Ferber, M., Richerson, D.W., Eds.; ASME PRESS: New York, NY, USA, 2003; pp. 1–10.
2. Steibel, J. Ceramic matrix composites taking flight at GE Aviation. *Am. Ceram. Soc. Bull.* **2019**, *98*, 30–33.
3. Jacobson, N.S. Corrosion of silicon-based ceramics in combustion environments. *J. Am. Ceram. Soc.* **1993**, *76*, 3–28. [CrossRef]
4. Opila, E.J.; Hann, R. Paralinear oxidation of CVD SiC in water vapor. *J. Am. Ceram. Soc.* **1997**, *80*, 197–205. [CrossRef]
5. Smialek, J.L.; Robinson, R.C.; Opila, E.J.; Fox, D.S.; Jacobson, N.S. SiC and Si_3N_4 scale volatility under combustor conditions. *Adv. Compos. Mater.* **1999**, *8*, 33–45. [CrossRef]
6. Robinson, R.C.; Smialek, J.L. SiC recession caused by SiO_2 scale volatility under combustion conditions: I. Experimental results and empirical model. *J. Am. Ceram. Soc.* **1999**, *82*, 1817–1825. [CrossRef]
7. Opila, E.J.; Smialek, J.L.; Robinson, R.C.; Fox, D.J.; Jacobson, N.S. SiC recession caused by SiO_2 scale volatility under combustion conditions: II. Thermodynamics and gaseous-diffusion model. *J. Am. Ceram. Soc.* **1999**, *82*, 1826–1834. [CrossRef]
8. More, K.L.; Tortorelli, P.F.; Ferber, M.K.; Walker, L.R.; Kiser, J.R.; Miriyala, N.; Brentnall, W.D.; Price, J.R. *ASME Paper 99-GT-292*; ASME TURBOEXPO: Indianapolis, IN, USA, 1999.
9. Ferber, M.K.; Lin, H.T.; Parthasarathy, V.; Brentnall, W.D. *ASME Paper 99-GT-265*; ASME TURBOEXPO: Indianapolis, IN, USA, 1999.
10. Opila, E.J.; Myers, D.L. Alumina volatility in water vapor at elevated temperatures. *J. Am. Ceram. Soc.* **2004**, *87*, 1701–1705. [CrossRef]
11. Lee, K.N.; Fritze, H.; Ogura, Y. *Progress in Ceramic Gas Turbine Development Vol. 2*; van Roode, M., Ferber, M., Richerson, D.W., Eds.; ASME PRESS: New York, NY, USA, 2003; pp. 641–664.
12. Lee, K.N. Environmental barrier coatings for CMCs. In *Ceramic Matrix Composites*; Bansal, N.P., Lamon, J., Eds.; Wiley: New York, NY, USA, 2015; pp. 430–451.
13. Lee, K.N. $Yb_2Si_2O_7$ environmental barrier coatings with reduced bond coat oxidation rates via chemical modifications for long life. *J. Am. Ceram. Soc.* **2019**, *102*, 1507–1521. [CrossRef]
14. Smialek, J.L. Relative Ti_2AlC scale volatility under 1300 °C combustion conditions. *Coatings* **2020**, *10*, 142. [CrossRef]
15. Vaßen, R.; Bakan, E.; Gatzen, C.; Kim, S.; Mack, D.E.; Guillon, O. Environmental barrier coatings made by different thermal spray technologies. *Coatings* **2019**, *9*, 784. [CrossRef]
16. Gatzen, C.; Mack, D.E.; Guillon, O.; Vaßen, R. $YAlO_3$-A novel environmental barrier boating for Al_2O_3/Al_2O_3–ceramic matrix composites. *Coatings* **2019**, *9*, 609. [CrossRef]
17. Klemm, H.; Schönfeld, K.; Kunz, W. Delayed formation of thermally grown oxide in environmental barrier coatings for non-oxide ceramic matrix composites. *Coatings* **2020**, *10*, 6. [CrossRef]
18. Kitaoka, S.; Matsudaira, T.; Kawashima, N.; Yokoe, D.; Kato, T.; Takata, M. Structural stabilization of mullite films exposed to oxygen potential gradients at high temperatures. *Coatings* **2019**, *9*, 630. [CrossRef]
19. Webster, R.U.; Bansal, N.P.; Salem, J.A.; Opila, E.J.; Wiesner, V.L. Characterization of thermochemical and thermomechanical properties of Eyjafjallajökull volcanic ash glass. *Coatings* **2020**, *10*, 100. [CrossRef]
20. Kabir, M.R.; Sirigiri, A.K.; Naraparaju, R.; Schulz, U. Flow kinetics of molten silicates through thermal barrier coating: A numerical study. *Coatings* **2019**, *9*, 332. [CrossRef]
21. Seo, H.-I.; Kim, D.; Lee, K.S. Crack healing in mullite-based EBC during thermal shock cycle. *Coatings* **2019**, *9*, 585. [CrossRef]
22. Kawai, E.; Kakisawa, H.; Kubo, A.; Yamaguchi, N.; Yokoi, T.; Akatsu, T.; Kitaoka, S.; Umeno, Y. Crack initiation criteria in EBC under thermal stress. *Coatings* **2019**, *9*, 697. [CrossRef]

Article

Relative Ti$_2$AlC Scale Volatility under 1300 °C Combustion Conditions

James L. Smialek [†]

NASA Glenn Research Center, Cleveland, OH 44135, USA; Dr.JSmialek@outlook.com
† Retired, Oct 29, 2018.

Received: 4 January 2020; Accepted: 3 February 2020; Published: 5 February 2020

Abstract: Turbine environments may degrade high temperature ceramics because of volatile hydroxide reaction products formed in water vapor. Accordingly, the volatility of transient TiO$_2$ and steady-state Al$_2$O$_3$ scales formed on the oxidation-resistant Ti$_2$AlC MAX phase ceramic was examined in 1300 °C high velocity (Mach 0.3, 100 m/s) and high pressure (6 atm, 25 m/s) burner rig tests (BRT). Unlike metals, the ceramic was stable at 1300 °C. Unlike SiC and Si$_3$N$_4$, neither burner test produced a weight loss, unless heavily pre-oxidized. Lower mass gains were produced in the BRT compared to furnace tests. The commonly observed initial, fast TiO$_2$ transient scale was preferentially removed in hot burner gas (~10% water vapor). A lesser degree of gradual Al$_2$O$_3$ volatilization occurred, indicated by grain boundary porosity and crystallographic etching. Modified cubic-linear (growth-volatility) kinetics are suggested. Gas velocity and water vapor pressure play specific roles for each scale. Furthermore, a 7YSZ TBC on Ti$_2$AlC survived for 500 h in the Mach 0.3 burner test at 1300 °C with no indication of volatility or spalling.

Keywords: MAX phases; scale volatility; burner rigs; EBC

---

1. Brief Introduction

Ceramics are widely indicated for higher temperature turbine components. Water vapor attack has become a new element of study because of volatile hydroxides that form by reaction with Al$_2$O$_3$ substrates or with SiO$_2$ scales that form on SiC and Si$_3$N$_4$. The severity of attack is controlled by the thermodynamics of the reaction, velocity, and water vapor pressure [1–4]. The water vapor component of combusted jet fuel is generally indicated at ~10%, with chemical activity increasing with overall system pressure. Typical volatile hydroxides of common oxides observed or projected are CrO$_2$(OH)$_2$, Si(OH)$_4$, TiO(OH)$_2$, and Al(OH)$_3$, in order of decreasing severity [1–4]. While Cr volatiles can occur at low temperatures, 1200 °C or more is required for the others to become a noticeable problem. Thus, widely used Al$_2$O$_3$-forming superalloys, typically limited to 1150 °C, have not previously exhibited a scale volatility problem. Also, thermal barrier coatings (TBC) allow for higher gas temperatures, but protect the metal and thermally grown oxide (TGO) from both high temperature and high velocity. Al$_2$O$_3$ and SiC ceramic matrix composites (CMC) generally need an environmental barrier coating (EBC) for continuous use above about 1200 °C [5]. Furthermore, SiO$_2$ scales are known to have significantly increased growth rates in water vapor, whereas Al$_2$O$_3$ scales show more complex effects on metals than on MAX phases [6–9].

Another class of materials that presents opportunities at intermediate temperatures is that of Al$_2$O$_3$-forming MAX phases, such as Cr$_2$AlC, Ti$_2$AlC, and Ti$_3$AlC$_2$. These compounds are very oxidation resistant, some at or above 1300 °C [10], although strength is lacking for unsupported, load-bearing use at this temperature. A broad program evolved at NASA Glenn to examine Type II low temperature hot corrosion resistance of Cr$_2$AlC coatings on superalloys, basic Al$_2$O$_3$ scale kinetics, and extreme durability of yttria-stabilized zirconia (YSZ) TBC on Ti$_2$AlC [11]. Burner rig testing was also performed that yielded experimental results for TiO_2 and Al_2O_3 scale durability in water vapor. The purpose of this present paper is to catalogue and analyze high pressure (6 atm) and high velocity (100 m/s) burner results with regards to scale volatility issues. Part of the motivation relates to the behavior of Al_2O_3 scales in high velocity water vapor at 1300 °C, an environment not typically used for standard metal alloy tests. The materials and processes were described in detail from a number of studies comprising the source data for this compilation [12–15]. All the results subsequently presented derive from these studies. For further reference, those studies included broad literature surveys and more in-depth discussions.

2. Materials and Methods

Ti_2AlC MAXthal 211® was obtained from Sandvik/Kanthal (Sandviken, Sweden) and EDM machined into ~0.2 cm thick ×1.2 cm wide ×2 cm long furnace samples, ×4 cm long HP-BRT samples, and ×7 cm long Mach 0.3 BRT samples. These were polished through 2400 grit and ultrasonically cleaned in detergent and alcohol. The box furnace tests (Rapid Temp) were conducted in lab air and samples intermittently weighed/inspected over graduated intervals for up to 500 h. Thermogravimetric (TGA) tests were conducted in dry, bottled air for 100 h, using a vertical tube furnace with weights continuously recorded by a thermo-balance (Setaram, Caluire, France). High pressure burner rig tests (6 atm, 25 m/s) were performed in a jet-fuel burner apparatus, completely sealed in a water-cooled stainless-steel chamber, with pressure controlled by an exhaust valve. Test bars were held at the ends in a water-cooled fixture, at 45° to the flame. Typical run series achieved 6 h exposure between shutdown and weightings, usually accruing 50 h total test time. Heating and cooldown were generally achieved within 10 min. High velocity Mach 0.3 BRT tests (1 atm., 100 m/s) employed an open jet-fueled burner, with a cantilever gripped, face-on test bar. The face had been coated with 160 μm of Metco 6700 7YSZ TBC by plasma spray-physical vapor deposition (PS-PVD, Sulzer-Oerlikon, 94 kW, 40/80 Ar/He, 1.5 mbar), the backside left uncoated. Cycling (5 h) was obtained by pivoting the burner away from the sample. Heating and cooling were generally complete in 1 min. BRT sample temperatures were monitored by two-color or 8 μm pyrometers, with gas temperature measured by thermocouple. Samples were examined at various intervals and weighed on a Sartorius analytical balance (±0.00005 g). Scales were examined by optical and FEG-SEM microscopy (Hitachi S-4700, Tokyo, Japan), both on the surface and as Ni-plated, metallographically polished cross-sections at the end of the 500 h BRT. Phase contents were estimated from (Brüker 8D) X-ray diffractometer scans and Rietveld refinement, using Jade software (version 6). Complete experimental details are available in the source studies [12–15].

3. Results

The weight change results from a suite of 1300 °C furnace and jet fuel burner tests are presented in Figure 1. Test times were dictated by the specific study they addressed: standard 100 h TGA, extended 300 h furnace pre-oxidation to produce a very slow growing Al_2O_3 scale, short 50 h HP-BRT to demonstrate kinetics in a labor-intensive, expensive, pressurized apparatus, and 500 h to demonstrate long term TBC durability in the convenient, available, Mach 0.3 high velocity cyclic rig.

The ambient air box furnace (pre-ox) and dry air TGA show similar gains. These are compared to lower curves for the two burner tests. All results indicate a rapid initial uptake in the order of 1 mg/cm^2 within the first hour of exposure. This has often been associated with a rapid transient growth of discontinuous TiO_2 scales that are then undercut by a healing layer of slow-growing, steady-state Al_2O_3. Nearly linear weight loss was observed in the HP-BRT for a pre-oxidized sample, that would otherwise have been masked by the high initial growth rate. Long-term testing was conveniently enabled by the box furnace, used for a pre-oxidation treatment here, and by the accessibility of the open Mach 0.3 burner test. Weight gain of ~2.4 mg/cm^2 was achieved in the Mach 0.3 BRT as compared to about 1.6 mg/cm^2 for the much shorter HP-BRT.

The cubic scaling kinetics were easily treated by correcting (subtracting) the amount of transient TiO_2, as graphically interpolated on log-log plots [12,14]. In the case of TGA tests, it was shown that most of the transient growth, w_0, took place in the first 10 min. Good linearized fits of $(w - w_0)$ to $(t - t_0)^{1/3}$ cubic kinetics could be obtained in these well-controlled TGA furnace tests. The cubic scaling constants, extracted from the $(w - w_0)$ offset-corrected mass gain curves show a very well-behaved, single-mechanism Arrhenius dependency, Figure 2. Over the temperature range of 1000 °C–1400 °C, an activation energy of 334 kJ/mol·K is seen to apply. Accordingly, the same graphical approach was applied to the 1300 °C test data for HP-BRT (Figure 3a) and Mach 0.3 BRT (Figure 3b). These yielded k_c of 0.024 and 0.011 mg^3/cm^6·h, respectively, compared to 0.212 mg^3/cm^6·h determined by TGA [12,13]. These BRT reductions reflect losses due to scale volatility effects. The Mach 0.3 test also incorporates protective effects of the YSZ face-coat and lower temperatures (100 °C) away from the hot zone.

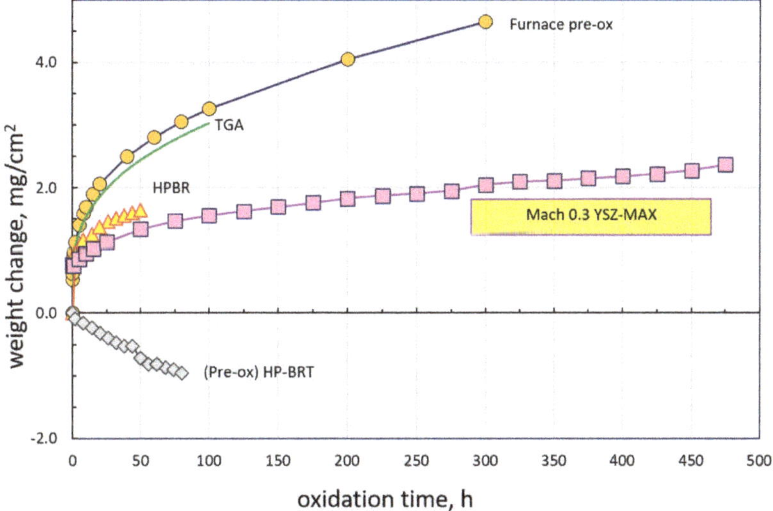

Figure 1. Comparison of 1300 °C Ti$_2$AlC furnace and burner oxidation data. Mach 0.3 burner at 1 atm. and 100 m/s; HPBR at 6 atm. and 25 m/s, TGA dry air, and ambient air furnace tests.

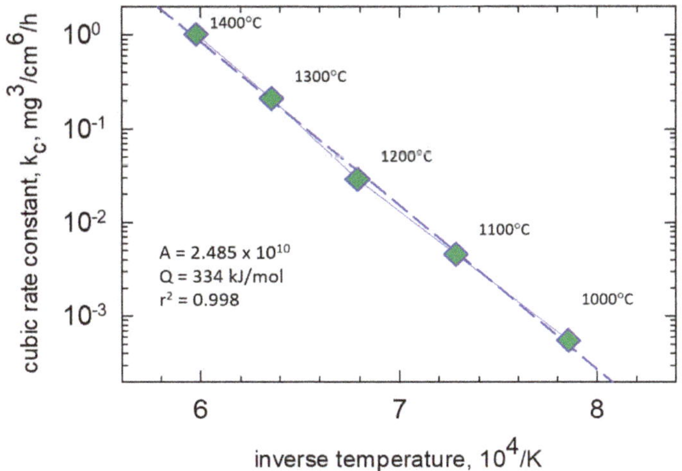

Figure 2. Arrhenius plot of log cubic oxidation rate constant vs 1/T; TGA furnace tests of Ti$_2$AlC in dry air for 100 h at 1000–1400 °C [14].

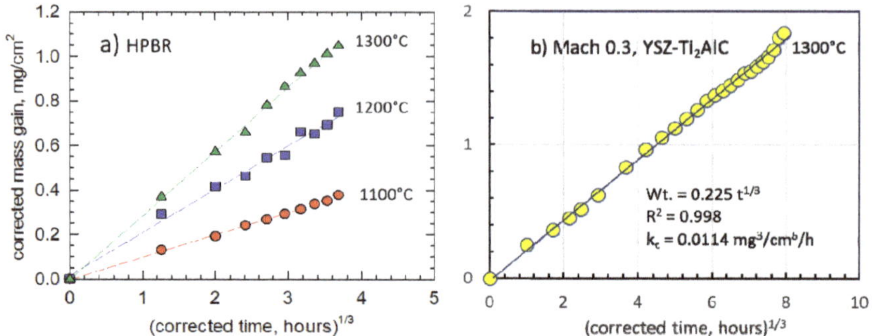

Figure 3. Cubic oxidation kinetics suggested by linearized transient-corrected weight vs $t^{1/3}$ behavior for Ti_2AlC. (**a**) 6 atm, 25 m/s, 50 h, HP-BRT; and (**b**) 1 atm., 100 m/s, 500 h, Mach 0.3 BRT, with YSZ face-coat [13,15].

Visual confirmation of TiO_2 volatility can be surmised from the low-magnification optical micrographs in Figure 4. Here, the scattered, sometimes oriented, initial clusters of light scale phases are seen to coarsen with furnace exposure time and decrease or disappear with HP-BRT exposure time. This effect was semi-quantitatively verified by Rietveld analyses of X-ray diffractometer scans. While TiO_2 (rutile) was the primary transient identified at 1200 °C and below, the reaction phase of $TiAl_2O_5$ was also identified after 1300 °C exposures. Here, initial scale quantities of 20% TiO_2 and 10% $TiAl_2O_5$ were determined after just 0.2 h of furnace exposure. These decreased dramatically to only 0.1% and 1%, respectively, after 80-h HP-BRT exposures (300 h furnace pre-oxidation), the remainder being α-Al_2O_3 [13].

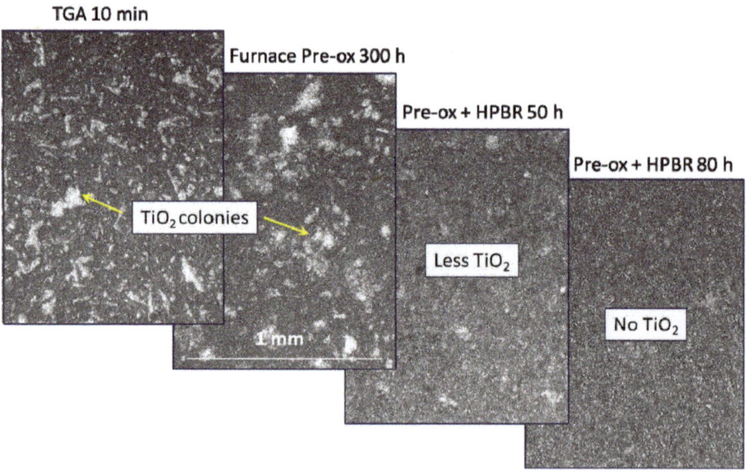

Figure 4. Optical micrographs depicting discontinuous TiO_2 scales formed in furnace tests and successive removal in HP-BRT exposures at 1300 °C.

The effect of this HP-BRT exposure on the surface structure can be seen in Figure 5. After 300 h pre-oxidation at 1300 °C, transient TiO_2 and $TiAl_2O_5$ bright clusters (T) were retained in (a), but then largely removed by HP-BRT testing for 80 h at 1300 °C in (b). Distinct underlying grains of Al_2O_3 (A) could then be discerned with a much lower Ti EDS signal overall. A linear weight loss rate of 0.012 mg/cm²·h was also measured, as shown by the lower curve in Figure 1. Since this included some modest scale growth, the total removal rate was surmised to be about 0.017 mg/cm^2·h. (Pre-oxidation was required to produce a thick scale with a low instantaneous growth rate less than the volatility rate).

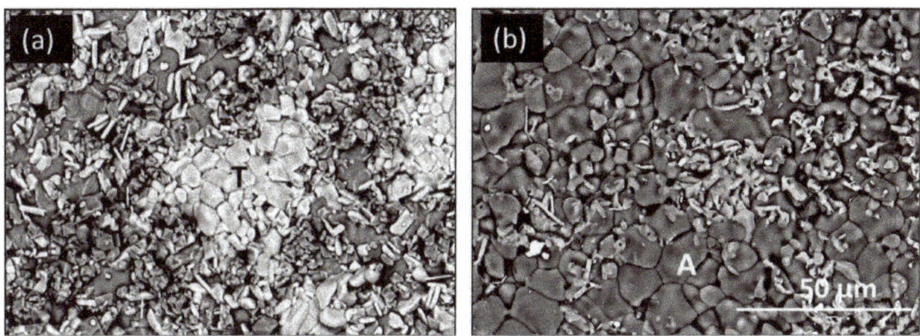

Figure 5. SEM/BSE surface microstructures of scales formed on Ti$_2$AlC at 1300 °C in (**a**) furnace pre-oxidation for 300 h and (**b**) followed by high pressure burner rig (80 h at 6 atm, 25 m/s). T (Ti-rich transient scale); A (Al$_2$O$_3$) [13].

A direct comparison of scales formed in 1300°C TGA (100 h) and HP-BRT (50 h) in cross-section is presented in Figure 6. The TGA structure shows the Ti-rich remnants of scattered transient scale colonies, with a dense underlayer of Al$_2$O$_3$. The HP-BRT sample exhibits a rather discontinuous surface scale with less distinct Ti-rich regions, if at all. HP-BRT scale volatility is again suggested. The inner Al$_2$O$_3$–Ti$_2$AlC interface is completely intact with no porosity or cracking.

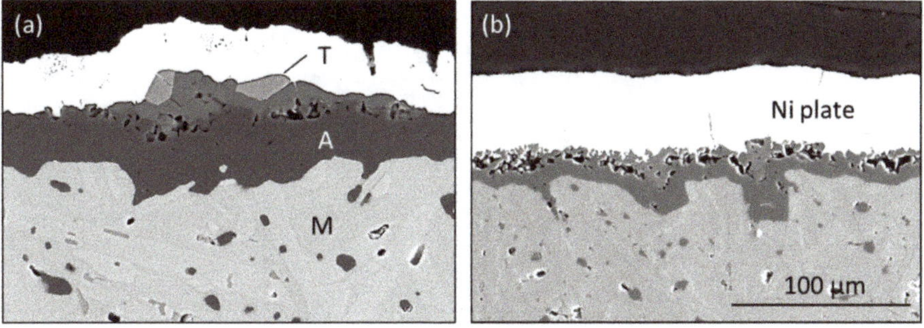

Figure 6. SEM/BSE cross sections of scale microstructures formed on Ti$_2$AlC at 1300 °C in (**a**) furnace TGA (100 h) [11] and (**b**) high pressure burner rig (50 h at 6 atm, 25 m/s). Ni (plating); T (Ti-rich transient scale); A (Al$_2$O$_3$); M (MAX phase substrate).

Mach 0.3 BRT (1 atm., 100 m/s, 500 h) exposures produced similar effects on surface scale microstructure (uncoated sample backside) (Figure 7). However, since the burner nozzle was about 2.5 cm in diameter centered on the 5 cm long exposed sample length, the sample temperature at the top (a) and bottom (grip end, c) was about 100°C cooler. This resulted in a less severe attack, with some remnants of bright Ti-rich particles atop Al$_2$O$_3$ grains, the latter exhibiting grain boundary porosity.

In contrast, the hot section (b) showed little vestige of Ti-rich scales, but a highly irregular, open Al$_2$O$_3$ scale. Some grains appeared to be etched crystallographically, forming lamellae, possibly along the hexagonal (0001) basal planes. The platelets retained a slight Ti level; they may have been derived from TiAl$_2$O$_5$ grains where Ti was removed by selective water vapor corrosion. Xrd analyses of the oxidized surface showed ~10% TiO$_2$ after the initial 20 min at 1000 °C, then removed by volatile reactions to just 0.1% after the 1300 °C exposure, the remainder of the scale being α-Al$_2$O$_3$ [15]. No phase change in the Ti$_2$AlC substrate was apparent other than reduced x-ray diffraction intensity due to absorption from the thickening scale.

In cross-section, Figure 8, little indication of Ti-rich scales remains, and the scale is thicker in the hot zone region (a). The surfaces are very irregular and open, consistent with scale removal by volatile products. Less of this structure remains in the hot zone region compared to the grip end (b). Again, the scale-substrate interface is seen to be completely intact.

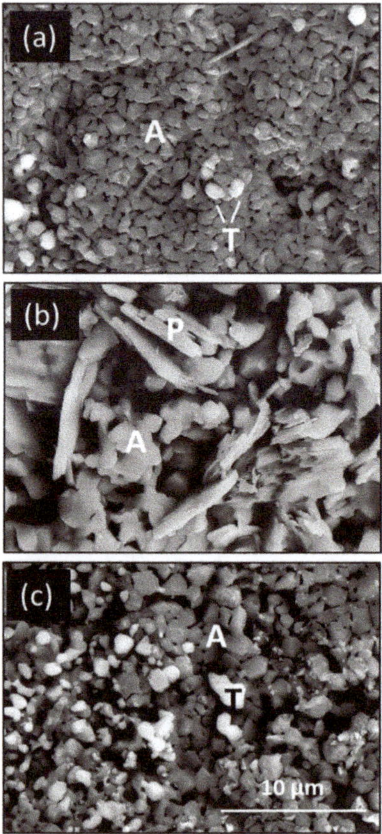

Figure 7. SEM/BSE surface microstructures of scales formed on Ti$_2$AlC at 1300 °C in Mach 0.3 high velocity burner rig (500 h at 1300 °C) Uncoated backside at sample (**a**) top, (**b**) hot center, (**c**) grip end. T (Ti-rich transient scale); A (Al$_2$O$_3$); P (etched platelets) [15].

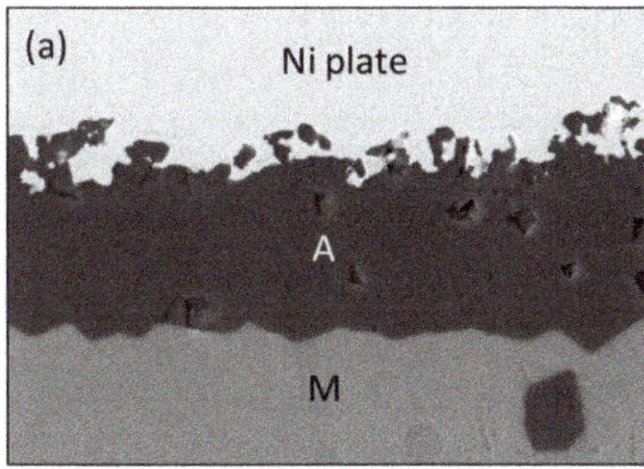
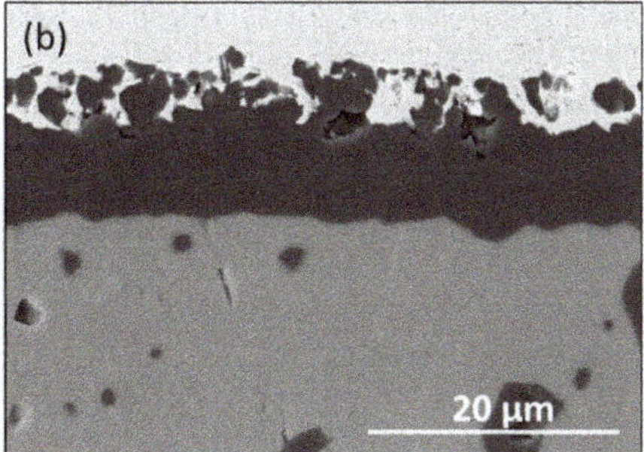

Figure 8. SEM/BSE cross-section images of scale microstructures formed on Ti$_2$AlC in Mach 0.3 BRT (1300 °C, 500 h, 100 m/s, 1 atm). (**a**) Hot section, uncoated backside; (**b**) cooler grip end, uncoated backside. Ni (plating); A (Al$_2$O$_3$); M (MAX phase substrate) [15].

Figure 9 presents coating structures typifying the as-sprayed (a) and the hot zone region (b) for the coated face of the Mach 0.3 BRT sample. The coating exhibits deposition columns, first textured by the deposition process (a), then by grain growth and surface smoothing during thermal annealing (b). Here, the flame directly impinged on the YSZ coating face, which shows no features of oxide removal by volatility: the zirconia grains and PS-PVD coating columns are basically intact. In the cross-section (Figure 10), the coating, TGO, and MAX phase substrate are also intact with no interfacial porosity, cracks, or delamination. Porosity and metallographic pullout, however, is observed within the scale. Volatility issues have therefore been prevented on the coated face. Coating survival after 500 h testing at 1300 °C and 100 m/s is thus indicated on all accounts. Further testing would not be "cost effective" or especially productive as there was little indication of imminent failure, i.e., the same results are expected for 1000 h testing and beyond. Furnace testing has shown similar durability for the YSZ/Ti$_2$AlC system, surviving 2500 h total, including 500 h at 1300 °C and scales up to 40 µm thick [14]. While little evidence is seen for detrimental interface reactions, it can be surmised that sustained Al$_2$O$_3$ growth will be limited by the Al reservoir in the MAX phase substrate.

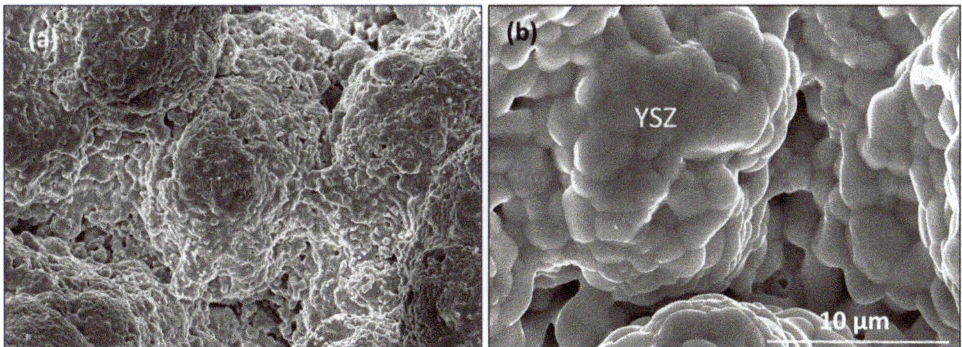

Figure 9. SEM microstructure of YSZ coating surface on Ti$_2$AlC after the Mach 0.3 high velocity burner rig test. (**a**) Grip end, 500 h/1200 °C test, (**b**) hot zone, 500 h/1300 °C test.

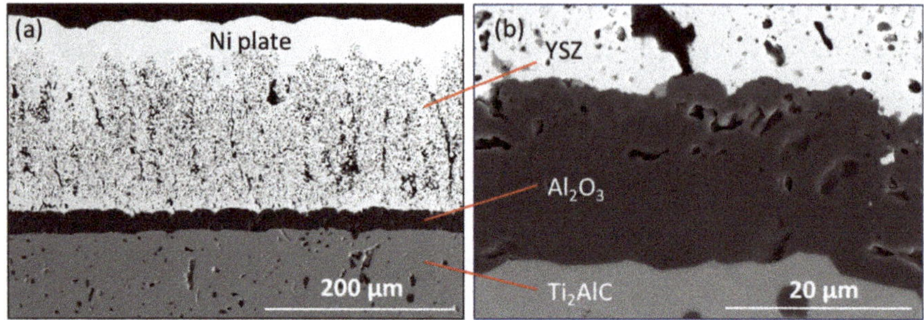

Figure 10. SEM/BSE hot zone cross-section of YSZ coated Ti_2AlC after BRT (Mach 0.3, 500 h/1300 °C). (**a**) Full structure, (**b**) detail showing intact YSZ–Al_2O_3–Ti_2AlC interfaces [15].

4. Discussion

The previous assemblage of results compared the high temperature scaling characteristics of the oxidation resistant Ti_2AlC phase under moist, high velocity burner conditions to those from static and dry atmospheres. The distinct appearance and removal of the initial TiO_2 transient scale by moisture in a high velocity gas stream was highlighted. The relative susceptibility of TiO_2 to $TiO(OH)_2$ formation in water vapor appears preferential compared to that of Al_2O_3 via $Al(OH)_3$ volatiles. Such a condition has been examined in concert with Jacobson's thermodynamic treatment of various oxides in flowing moist gases [13]. It was predicted that $TiO(OH)_2$ losses would be on the same order as $Al(OH)_3$ losses, but it is now acknowledged that some uncertainties still remain regarding TiO_2 volatiles [16,17]. While the Ti-oxides appeared to be removed preferentially, some losses of Al_2O_3 are also indicated. Critical studies have indeed demonstrated volatile losses and crystallographic etching for bulk Al_2O_3 [18–20].

The general removal rate of various scales in various water vapor environments can be modeled according to $v^{1/2}p_{H2O}{}^n/p_{total}{}^{1/2}$, using the original thermochemical-diffusional approach developed by Opila et al. for various oxides [1]. Here, $n = 1$ for TiO_2, $n = 3/2$ for Al_2O_3, and $n = 2$ for SiO_2, as dictated by the chemical reaction with water vapor. Accordingly, the relative severity of the Mach 0.3 test (100 m/s, 1 atm) to the HP-BRT (25 m/s, 6 atm) shown here produces relative rig factors of 0.82, 0.33, and 0.14 for the three scales, respectively [15]. Thus, in the Mach 0.3 test, TiO_2 is expected to show similar attack severity as in the HP-BRT, while SiO_2 is expected to show less attack, with Al_2O_3 intermediate. This is consistent with efficient removal of TiO_2 observed in both tests and more severe removal of SiO_2 in the HP-BRT.

An attempt was made to extract volatility kinetics from the weight change curves using a cubic-linear fit (Chen-Tedmon) [21]. This was partially enabled using COSP for Windows originally designed for cyclic oxidation spalling models. For the present case, cubic growth and uniform scale removal is the model selected [13,22]. To account for a decreasing amount of TiO_2 with time, the "spall" (removed) thickness exponent (α) is addressed as a negative number (-3). It is recognized that a constant quotient of the "spall" fraction, Q_0, and cycle duration, τ, yield identical loss rates per hour. (The response basically converges to continuous curves for $\tau \leq 1$ h and reproduce their analytical expression).

Some solutions are presented in Figure 11 for (a) the HP-BRT and (b) Mach 0.3 BRT results. Both model curves (dashed) are reasonable fits for the experimental data (symbols). The fitting parameters were $k_c = 0.212$ mg^3/cm^6/h, $Q_0 = 0.220$ for the HP-BRT [13] and $k_c = 0.050$ mg^3/cm^6/h, $Q_0 = 0.038$ for the Mach 0.3 BRT. Both sets fixed a cubic growth exponent, $m = 3.0$, and used a decreasing spall exponent, $\alpha = -3$. The HP-BRT fit initiated with the same k_c determined by TGA in dry air. The resulting Q_0 was shown to be consistent with a linear volatility weight loss of 0.01–0.02 mg/cm^2·h produced for the pre-oxidized sample [13]. The Mach 0.3 BRT, however, maintained face-to-back and center-to-top/bottom temperature gradients. Accordingly, the net Mach 0.3 growth constant was considerably less, being only about one-quarter that of the HP-BRT. Also, the COSP fit projected a low Al volatility weight loss rate of ~0.002 mg/cm^2·h after 100 h, converging to just ~0.001 mg/cm^2·h after 500 h, or about one-tenth that of the 50 h HP-BRT rates. Remarkably, the Al$_2$O$_3$ scale thickness under the YSZ in the hot zone (at ~1244 °C) after 500 h (~20 μm) was basically the same as that produced in the HP-BRT (at ~1300 °C) after just 50 h.

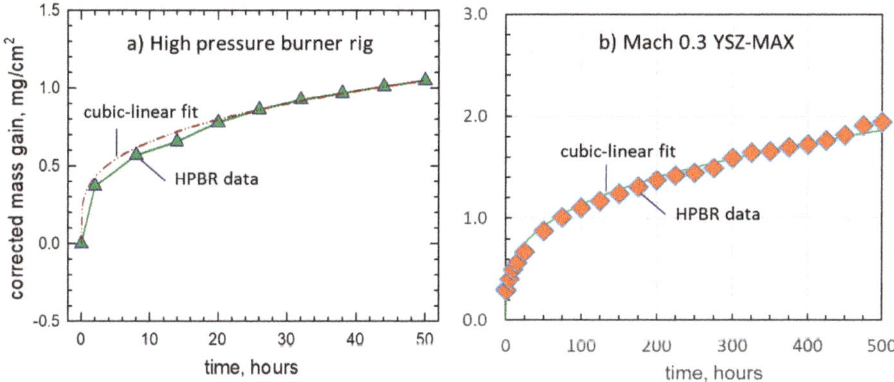

Figure 11. Cubic-linear fits to burner oxidation results for Ti$_2$AlC: (**a**) high pressure (6 atm, 25 m/s, 1300 °C). (COSP for Windows, k_c = 0.212 mg^3/cm^6/h, Q = 0.220 mg/cm^2/h, m = 3, α = −3); (**b**) high velocity Mach 0.3 for YSZ/Ti$_2$AlC (1 atm, 100 m/s, 1300 °C). (COSP for Windows: k_c = 0.050 mg^3/cm^6/h, Q = 0.038 mg/cm^2/h, m = 3, α = −3).

5. Summary

Results from various thermal exposures of Ti$_2$AlC incorporating water vapor attack in high velocity gas have been examined. Volatility of TiO$_2$ and Al$_2$O$_3$ scales at 1300 °C was indicated by the burner rig results, especially when compared to furnace tests. Discontinuous, superficial colonies of Ti-rich oxide grains were essentially cleaned off in both high-pressure (6 atm.) and long-term, high-velocity (100 m/s) burner tests. While moderate weight gains resulted from continuous Al$_2$O$_3$ growth, 300 h pre-oxidation allowed a net weight loss to be observed for a thick slow-growing scale. The linear loss rate was about one-fifth that determined for SiO$_2$ scales formed on SiC in the same exposure. While Al$_2$O$_3$ losses were not as pronounced as TiO$_2$, grain boundary etching and porosity indicated some volatility effects. The PS-PVD YSZ coating on the hot face of the Mach 0.3 test sample showed no volatility effects and was totally protective of the underlying adherent Al$_2$O$_3$ scale and Ti$_2$AlC substrate. No degradation was apparent for 500 h at a temperature (1300 °C) well in excess of current alloy system capabilities. Continued testing was unwarranted and impractical since failure would not be expected even for much longer times.

Funding: This research received no external funding.

Acknowledgments: Substantial contributions were made to the original works by M. Cuy, B. Harder, A. Garg, R. Pastel, J. Buehler, R. Rogers, and N. Jacobson. Those studies were performed under the NASA Fundamental Aeronautics Program.

Conflicts of Interest: The author declares no conflict of interest

References

1. Opila, E.J. Volatility of common protective oxides in high-temperature water vapor: Current understanding and unanswered questions. *Mater. Sci. Forum* **2004**, *461*, 765–774. [CrossRef]
2. Jacobson, N.; Myers, D.; Opila, E.; Copland, E. Interactions of water vapor with oxides at elevated temperatures. *J. Phys. Chem. Solids* **2005**, *66*, 471–478. [CrossRef]

3. Opila, E.J.; Jacobson, N.S.; Myers, D.L.; Copland, E.H. Predicting oxide stability in high-temperature water vapor. *JOM* **2006**, *58*, 22–27. [CrossRef]

4. Meschter, P.J.; Opila, E.J.; Jacobson, N.S. Water vapor–mediated volatilization of high-temperature materials. *Annu. Rev. Mater. Res.* **2013**, *43*, 559–588. [CrossRef]

5. Lee, K.N.; Fox, D.S.; Bansal, N.P. Rare earth silicate environmental barrier coatings for SiC/SiC composites and Si3N4 ceramics. *J. Eur. Ceram. Soc.* **2005**, *25*, 1705–1715. [CrossRef]

6. Basu, S.; Obando, N.; Gowdy, A.; Karaman, I.; Radovic, M. Long-term oxidation of Ti_2AlC in air and water vapor at 1000 °C–1300 °C temperature range. *J. Electrochem. Soc.* **2012**, *159*, C90. [CrossRef]

7. Maris-sida, M.C.; Meier, G.H.; Pettit, F.S. Some water vapor effects during the oxidation of alloys that are α-Al_2O_3 formers. *Metall. Mater. Trans. A* **2003**, *34A*, 2609–2619. [CrossRef]

8. Lin, Z.J.; Li, M.S.; Wang, J.Y.; Zhou, Y.C. Influence of water vapor on the oxidation behavior of Ti_3AlC_2 and Ti_2AlC. *Scr. Mater.* **2008**, *58*, 29–32. [CrossRef]

9. Unocic, K.A.; Pint, B.A. Effect of water vapor on thermally grown alumina scales on bond coatings. *Surf. Coat. Technol.* **2013**, *215*, 30–38. [CrossRef]

10. Tallman, D.J.; Anasori, B.; Barsoum, M.W. A critical review of the oxidation of Ti_2AlC, Ti_3AlC_2 and Cr_2AlC in air. *Mater. Res. Lett.* **2013**, *1*, 115–125. [CrossRef]

11. Smialek, J.L. Oxidation of Al_2O_3 scale-forming MAX phases in turbine environments. *Metall. Mater. Trans. A* **2017**, *49A*, 782–792. [CrossRef]

12. Smialek, J.L. Kinetic aspects of Ti_2AlC MAX phase oxidation. *Oxid. Met.* **2015**, *83*, 351–366. [CrossRef]

13. Smialek, J.L. Environmental resistance of a Ti_2AlC-type MAX phase in a high pressure burner rig. *J. Eur. Ceram. Soc.* **2017**, *37*, 23–34. [CrossRef]

14. Smialek, J.L.; Harder, B.J.; Garg, A. Oxidative durability of TBCs on Ti_2AlC MAX phase substrates. *Surf. Coat. Technol.* **2016**, *285*, 77–86. [CrossRef]

15. Smialek, J.L.; Cuy, M.D.; Harder, B.J.; Garg, A.; Rogers, R.B. Durability of YSZ Coated Ti2AlC in 1300°C Mach 0.3 Burner Rig Tests; NASA/TM 2020-220380. 2019. Available online: https://ntrs.nasa.gov/search.jsp?R=20190033077 (accessed on 4 February 2020).

16. Nguyen, Q.N.; Bauschlicher, C.W.; Myers, D.L.; Jacobson, N.S.; Opila, E.J. Computational and experimental study of thermodynamics of the reaction of titania and water at high temperatures. *J. Phys. Chem. A* **2017**, *121*, 9508–9517. [CrossRef]

17. Myers, D.L.; Jacobson, N.S.; Bauschlicher, C.W.; Opila, E.J. Thermochemistry of volatile metal hydroxides and oxyhydroxides at elevated temperatures. *J. Mater. Res.* **2019**, *34*, 397–407. [CrossRef]

18. Opila, E.J.; Myers, D.L. Alumina volatility in water vapor at elevated temperatures. *J. Am. Ceram. Soc.* **2004**, *87*, 1701–1705. [CrossRef]

19. Yuri, I.; Hisamatsu, T. Recession Rate Prediction for Ceramic Materials in Combustion Gas Flow. 2003. Available online: https://asmedigitalcollection.asme.org/GT/proceedings-abstract/GT2003/36843/633/298920 (accessed on 4 February 2020).

20. Gatzen, C.; Mack, D.E.; Guillon, O.; Vaßen, R. Water vapor corrosion test using supersonic gas velocities. *J. Am. Ceram. Soc.* **2019**, *102*, 6850–6862. [CrossRef]

21. Chen, Y.; Tan, T.; Chen, H. Oxidation companied by scale removal: Initial and asymptotical kinetics. *J. Nucl. Sci. Technol.* **2008**, *45*, 662–667. [CrossRef]

22. Smialek, J.L.; Auping, J. V COSP for windows—Strategies for rapid analyses of cyclic-oxidation behavior. *Oxid. Met.* **2002**, *57*, 559–581. [CrossRef]

Article

Environmental Barrier Coatings Made by Different Thermal Spray Technologies

Robert Vaßen [1,*], **Emine Bakan** [1], **Caren Gatzen** [1], **Seongwong Kim** [1,2], **Daniel Emil Mack** [1] and **Olivier Guillon** [1,3]

[1] Forschungszentrum Jülich GmbH, IEK-1, 52425 Jülich, Germany; e.bakan@fz-juelich.de (E.B.); c.gatzen@fz-juelich.de (C.G.); woods3@kicet.re.kr (S.K.); d.e.mack@fz-juelich.de (D.E.M.); o.guillon@fz-juelich.de (O.G.)
[2] Korea Institute of Ceramic Engineering and Technology (KICET), Engineering Ceramics Center, Shindun-myeon, Icheon-si, Gyeonggi-do 17303, Korea
[3] Jülich Aachen Research Alliance: JARA-Energy, Jülich 52425, Germany
* Correspondence: r.vassen@fz-juelich.de

Received: 6 October 2019; Accepted: 18 November 2019; Published: 22 November 2019

Abstract: Environmental barrier coatings (EBCs) are essential to protect ceramic matrix composites against water vapor recession in typical gas turbine environments. Both oxide and non-oxide-based ceramic matrix composites (CMCs) need such coatings as they show only a limited stability. As the thermal expansion coefficients are quite different between the two CMCs, the suitable EBC materials for both applications are different. In the paper examples of EBCs for both types of CMCs are presented. In case of EBCs for oxide-based CMCs, the limited strength of the CMC leads to damage of the surface if standard grit-blasting techniques are used. Only in the case of oxide-based CMCs different processes as laser ablation have been used to optimize the surface topography. Another result for many EBCs for oxide-based CMC is the possibility to deposit them by standard atmospheric plasma spraying (APS) as crystalline coatings. Hence, in case of these coatings only the APS process will be described. For the EBCs for non-oxide CMCs the state-of-the-art materials are rare earth or yttrium silicates. Here the major challenge is to obtain dense and crystalline coatings. While for the Y_2SiO_5 a promising microstructure could be obtained by a heat-treatment of an APS coating, this was not the case for $Yb_2Si_2O_7$. Here also other thermal spray processes as high velocity oxygen fuel (HVOF), suspension plasma spraying (SPS), and very low-pressure plasma spraying (VLPPS) are used and the results described mainly with respect to crystallinity and porosity.

Keywords: environmental barrier coatings; thermal spray methods; atmospheric plasma spraying; suspension plasma spraying; very low-pressure plasma spraying; high velocity oxygen fuel spraying

1. Introduction

Ceramic materials often show unique high temperature capability. However, monolithic ceramics suffer from an intrinsic low fracture toughness and, hence, for demanding high temperature applications as blades, vanes, shrouds or transition ducts of turbine engines reinforced ceramics have to be used [1]. Even sensitive parts can be made from such ceramic matrix composites (CMCs) and this has been demonstrated in the last years by different industries especially by GE [2]. For applications with higher mechanical loads, typically, non oxide CMCs as SiC/SiC are used due to their excellent high temperature strength and creep properties. However, also oxide/oxide composites which are mainly based on alumina and typically have reduced high temperature properties are of interest for specific applications e.g., combustion chamber applications [3]. In contrast to their metallic counterparts, CMCs do not have to be protected against oxidation as SiC forms a well-protecting silica scale and the oxide-based ceramics are anyhow stable at high temperatures in oxidizing atmosphere. However, these oxides tend to be rather unstable in the gas turbine atmosphere, namely water vapour containing combustion gases having velocities above 100 m/s and temperatures above 1100 °C. In this environment volatile (oxy) hydroxides are formed which remove the scale and can lead to a massive damage of the hole component [4,5]:

$$SiO_2 + 2\,H_2O \rightarrow Si(OH)_4 \uparrow \text{ or } Al_2O_3 + 3\,H_2O \rightarrow 2\,Al(OH)_3\uparrow \tag{1}$$

This water vapor recession can be addressed by coatings using highly corrosion-resistant materials (so-called environmental barrier coatings (EBCs)). Many interesting materials have been tested in high velocity and high-temperature and water vapor environments. The recession rates and coefficients of thermal expansion of most of the EBC materials and CMCs of interest in this study are summarized in Figure 1. The recession rate and the thermal expansion coefficient (TEC) is plotted for different materials and the values for two CMC materials are given, an SiC/SiC and an alumina-based Ox/Ox composite. Additionally, $Gd_2Zr_2O_7$ (CTE = $10.6 \times 10^{-6}\cdot K^{-1}$) was considered and was shown to have the lowest recession rate among pyrochlore materials successfully used for thermal barrier coating (TBC) application [6,7]. Generally, suitable EBC materials should have low corrosion rates and TECs close to the used CMC.

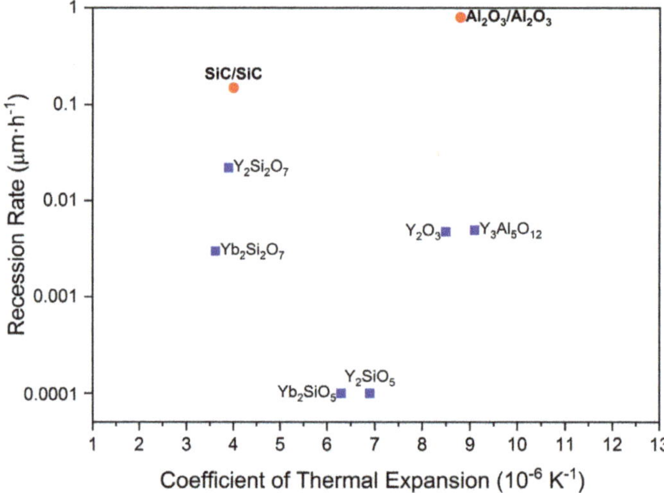

Figure 1. Recession rate and thermal expansion coefficient for EBC materials relevant to this study (blue) and the SiC/SiC and Al_2O_3/Al_2O_3 composites (red) (Data from [8–10]).

Compared to thermal barrier coatings (TBCs) EBCs should have a low porosity level to avoid the access of water vapour to the CMC. This will lead to rather high Young's moduli E_{EBC} in the EBCs close to the bulk values. A high Young's modulus is not favourable with respect to failure of the coatings as it increases the stored elastic energy; if this so-called energy release rate G exceeds a critical value G_c, a crack can propagate and failure occurs. The energy release rate can be calculated under plane stress conditions for a homogeneous coating stress σ and an infinite crack as:

$$G = \frac{\sigma^2 h}{2E_{EBC}} \tag{2}$$

with h being the coating thickness. If the stress in the coating relaxes at high temperature, the stress at room temperature is for a thin coating determined by the mismatch of thermal expansion coefficients:

$$\sigma = E_{EBC}(\alpha_{EBC} - \alpha_{CMC})\, \Delta T \tag{3}$$

As the square of TEC mismatch is determining the energy release rate in (2) a large mismatch should be avoided. Thickness and Young´s modulus have a linear influence on it.

The most often used technology to apply EBCs is thermal spray. Thermal spray is a versatile coating process in which the feedstock material is accelerated by a process gas and in most cases heated in powder form and then deposited on the substrate [11]. In thermal spray processes, the bonding between coating and substrate is mainly due to mechanical interlocking at asperities of the substrate surface. A sufficient bonding is typically achieved by a grit-blasting process prior to the deposition. However, it was found that for low strength materials as oxide-based composites this treatment can lead to significant degradation of the material. Hence, new methods as laser ablation to structure the surface will be discussed in the present paper.

In the thermal spray process, the molten particles will impinge on the substrate, the deformed droplets ("splats") will cool down quickly and generate large tensile stresses which typically introduce cracks [12]. These cracks form an open network which allow the diffusion of gas species through it and hence, such coatings will delay the water vapor corrosion but can hardly avoid it. In order to form gas-tight coatings rather hot spraying conditions with high substrate temperatures (above 500°C) and complete melting of the feedstock are beneficial. Such conditions can lead to a re-melting of already deposited splats and by this improve the bonding in the coating. Such approaches have been used to produce gas-tight electrolytes with a thickness below 50 μm from yttria stabilized zirconia (YSZ) for solid oxide fuel cells [13]. It was also demonstrated that high particle velocities at impact (>300 m/s) are favourable. Similar process conditions are also used to deposit segmented TBCs. In this process very dense coatings are produced by APS leading also to high tensile stresses in the coatings [14]. If a certain thickness (typically about 100 μm) is reached, the energy release rate exceeds the critical one and segmentation cracks are formed. This is of course not wanted for EBC applications. The segmentation cracks are driven by the cooling of the splats and hence by the absolute TEC of the ceramic coating material and not by the mismatch to the substrate. As a result EBCs for SiC/SiC should be less prone to segmentation cracks then those for Ox/Ox CMCs assuming the same other materials properties (Young´s modulus, toughness).

In addition to the difficulty to produce very dense coatings another issue is often important for EBCs. The materials used tend to form an amorphous phase when cooled down quickly from the molten state to the substrate temperature. This is detrimental, as the coatings will typically crystallize during operation, which is accompanied by a volume reduction and pore/crack formation in the coating. So, a highly crystalline coating in the as-sprayed condition is desirable. Previous results indicate that high deposition temperatures are favourable as then the time at temperature for crystallization is sufficient. It was proposed to deposit coatings in a furnace [15], however, we try in our approaches to avoid it as it introduces additional process steps and hence costs.

For the deposition of the EBCs we have investigated four different thermal spray processes (see Figure 2). Basic requirements for the processes are the capability to reach high feedstock temperatures and velocities.

First of all, atmospheric plasma spraying (APS) will be considered as a kind of standard baseline process. Here a fast variant (high velocity APS, HV-APS) will be described. The APS process shows a high temperature of the process gases (>10000 K at the nozzle exit) due to the efficient heating by electric arcs. Hence, a good and fast heat-transfer to the particles is possible, which allows complete particle melting, however can also lead to loss of those constituents in the powder with high vapor pressure. Typical velocities are in the range of 200–300 m/s, however specific conditions as small nozzle diameter and larger amounts of process gas allow higher velocities (up to 500 m/s, HV-APS)

While in the APS process particle in the range of several 10 micrometers (see splat in Figure 2) are used, the suspension plasma spraying allows much finer droplets and splats due to the use of suspensions. Here the droplet size is determined by a complex atomization process. The droplets typically in the size range of micrometers can easily reach the temperature and velocities of the process gas, hence both are typically higher than in APS. If the right suspension properties are chosen this can lead to improved densities, however also some more significant loss of volatile species can be expected due to the higher surface area of the droplets.

The plasma guns can also be operated in inert gas environment. For that a vacuum chamber is evacuated and then refilled by inert gas to typical pressures of about 50 mbar. The process is often. used to deposit metals to avoid in-flight oxidation. Due to the expansion of the plume into a chamber with reduced pressure, the gas velocities are higher than in APS. In our used equipment, powerful pumping units allow to maintain even lower chamber pressures in the mbar range during operation (Very low-pressure plasma spraying-VLPPS). This leads to even higher gas and particle velocities (see the large expanding plume in Figure 2) with corresponding high coating densities. In addition, the used powerful gun and the reduced cooling by convection allows homogenous high substrate temperatures which support dense and crystalline coating formation. This could be demonstrated for the deposition of dense ceramic membranes [16].

The last process which will be considered is a combustion process, the high velocity oxygen fuel (HVOF) spraying. The maximum gas temperature is limited to temperatures slightly above 3000 °C, so the heating of the feedstock is limited. On the other hand, the use of a de Laval nozzle allows extremely high, supersonic gas velocities (see shock diamonds in Figure 2) and corresponding high particle velocities. With such high velocities dense coating might be possible even without complete melting of the feedstock. Such a particle with non-molten core is visible in Figure 2

An additional more detailed description of some of the results can be found in our previous papers [17–20].

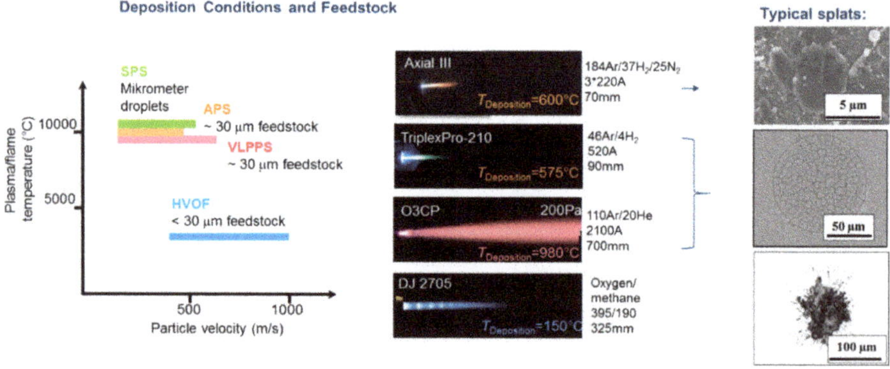

Figure 2. Characteristics of the 4 investigated thermal spray processes (Axial III used for SPS, TriplexPro210 for APS, O3CP for VLPPS, Dj2700 for HVOF) showing plume temperatures and particle velocities, photos of the plumes plus conditions used (process gases, current in case of plasma processes, and stand-off distances), and photos of typical splats.

2. Materials and Methods

2.1. Feedstock Materials, Suspensions and Substrates

Different feedstock powders have been used for the investigations, Table 1 gives an overview of the used materials.

Table 1. Description of the used feedstock materials.

Material	Manufacturer	Mean particle Size d_{50} [mm]	Process Used	Impurity Phases [wt.%]
Y_2SiO_5-YS	Treibacher Industrie AG	34	APS	7% Y_2O_3, 5% $Y_2Si_2O_7$
$Yb_2Si_2O_7$-YDS	Oerlikon Metco (US) Inc.	49	APS/VLPPS HVOF	5% Yb_2SiO_5
$Gd_2Zr_2O_7$	IEK-1	<30	APS	–
$Y_3Al_5O_{12}$	Treibacher	30	APS	–
Y_2O_3	Oerlicon Metco Europe GmbH	32	APS	–
$Yb_2Si_2O_7$-YDS	IEK-1	3.1	SPS	5% Yb_2SiO_5

For SPS purposes, YDS powder was milled in ethanol with the addition of polyethylenimine (PEI) and zirconia milling balls (d = 3 mm). The mixture was milled on roller cylinder (60 min^{-1}, 48 h) in order to produce a homogeneously-dispersed suspension (30 wt % in solid content). After milling, the suspension was diluted with ethanol to 10 wt % of solid loading and the particle size distribution was $d10$ = 1.3 µm, $d50$ = 3.1 µm, and $d90$ = 5.0 µm.

As substrates different materials have been taken as carbon steel, graphite, monolithic SiC and partly CMCs as 2d C/SiC (Schunk CF 226 P 75), SiC/SiCN (provided by DLR Stuttgart [21]) and an alumina based CMC FW12 (Walter E. C. Pritzkow Spezialkeramik, Filderstadt-Sielmingen, [22]). In the case of YS, a bond-coated IN738 was used as substrate for preliminary tests.

2.2. Thermal Spray Facilities

At IEK-1 different thermal spray systems for the deposition of ceramic coatings with high particle velocities are available. The "work horse" is a MultiCoat system (Oerlikon Metco, Wohlen, Switzerland) which was operated in this study for APS with the three-cathode TriplexPro 210 and for HVOF with the Diamond Jet 2700 spray torches both mounted on a six-axis robot (IRB 2400, ABB, Zurich, Switzerland). For the suspension plasma spraying a different spray booth equipped with the Axial III plasma torch with three separate cathode-anode pairs (Mettech Northwestern Corp., North Vancouver, BC, Canada) which allowed the axial injection of the liquid feedstock was used. As feeding system a home-made device was employed.

Vacuum plasma spraying was performed in a MultiCoat platform (Oerlikon Metco, Wohlen, Switzerland) with a 6 m^3 tank volume. Here the powerful O3CP torch (Oerlikon Metco, Wohlen, Switzerland) was chosen as plasma torch. The deposition temperatures were monitored by infrared pyrometers in each processing method.

An overview of the employed spray parameters is found in Table 2. The parameter used are based on long-year experience in the thermal spray of ceramics. Some further optimization trials have been made especially for EBC materials for oxide-based CMCs but they are not shown here. As shown in the results, the coating microstructure appears rather dense and often crystalline, no efforts to further optimize the microstructure were made. In the case of YDS the APS coating properties were not satisfying, here also other techniques have been investigated.

2.3. Surface Treatment

All samples were ultrasonically cleaned before spraying. To increase the adhesion properties, some of the Al_2O_3/Al_2O_3 CMC samples were pre-treated mechanically or by laser ablation. The mechanical pre-treatment was carried out with 80-grit sandpaper or by grit blasting with F 36 powder.

Table 2. Overview of the used thermal spray conditions (YS = Y_2SiO_5, YBDS = $Yb_2Si_2O_7$)

Process	APS	HVOF	VLPPS	SPS
Torch	Triplex Pro 210	DJ2700	O3CP	Axial III
Process gases [slpm]	YS:46Ar/4H$_2$ or 4He, YDS: 46Ar/4H$_2$ Gd$_2$Zr$_2$O$_7$: 46 Ar/4He Yb$_2$Si$_2$O$_7$: 46 Ar/4He Y$_3$Al$_5$O$_{12}$: 50 Ar/4He Y$_2$O$_3$: 46 Ar/4He	YDS: O$_2$ 395/CH$_4$ 190	YDS: Ar 110/He 20	245 in total [Ar/N$_2$/H$_2$] = 75%:10%:15%
Power[kW]	YS: He: 50, H$_2$: 58, YDS: 57 Gd$_2$Zr$_2$O$_7$: 50Yb$_2$Si$_2$O$_7$: 49.7 Y$_3$Al$_5$O$_{12}$: 44.4 Y$_2$O$_3$: 50.8	–	YDS: 90	YDS: 89
Stand-off distance [mm]	YS: 90/120 YDS: 90 Gd$_2$Zr$_2$O$_7$: 120 Yb$_2$Si$_2$O$_7$: 120 Y$_3$Al$_5$O$_{12}$: 70 Y$_2$O$_3$: 120	YDS: 325	YDS: 700	YDS: 50/70/90
Substrate temperature [°C]	YS: 460-750 YDS: 550 Gd$_2$Zr$_2$O$_7$: 500 Yb$_2$Si$_2$O$_7$: 600 Y$_3$Al$_5$O$_{12}$: 700 Y$_2$O$_3$: 500	YDS:150	YDS: 1000	YDS: 720/650/620
Substrates	YS: IN738, C/SiC, YDS: SiC Gd$_2$Zr$_2$O$_7$: Al$_2$O$_3$/Al$_2$O$_3$ Yb$_2$Si$_2$O$_7$: Al$_2$O$_3$/Al$_2$O$_3$ Y$_3$Al$_5$O$_{12}$: Al$_2$O$_3$/Al$_2$O$_3$ Y$_2$O$_3$: Al$_2$O$_3$/Al$_2$O$_3$	YDS: Steel and SiC/SiCN	YDS: Steel and SiC/SiCN	Monolithic SiC

A pulsed laser Trumpf TruMark 5020 with a wavelength of 1062 nm (Nd:YAG), 50 ns pulse duration and a maximum peak power of 15 kW was used for surface structuring (Trumpf GmbH + Co. KG, Ditzingen, Germany). The used laser parameters are given in Table 3. For a more detailed description of the ablation process the reader is referred to Gatzen et al. [23]

Table 3. Used laser parameters for surface structuring of Al_2O_3/Al_2O_3 CMCs.

Structure	Pulses Per Point	Program Repetition	Dot Distance (μm)
Honeycomb	1	50	30
Cauliflower	50	1	30

2.4. Characterization

After spraying the samples were sectioned, polished, and examined with a scanning electron microscope (Carl Zeiss NTS GmbH, Oberkochen, Germany) combined with an energy-dispersive

X-ray INCAEnergy355 spectrometer (EDS, Oxford Instruments Ltd., Abingdon, Oxfordshire, UK). Alternatively, SEM images were taken with a Hitachi TM3000 (Hitachi, Krefeld, Germany). Acquired SEM images were employed to assess the porosity in the coatings by means of image analysis using an image thresholding procedure with the analysis pro software (Olympus Soft Imaging Solutions GmbH, Germany). The analysis was performed on 10 SEM micrographs (2000× magnification) per sample, each with a resolution of 1280 × 1100 pixels and covering a horizontal field width of 126 μm. Crack density values of samples manufactured by suspension plasma spraying were calculated from the number of vertical cracks by using 15 SEM images (with ×300 magnification and width of 600 μm) as well.

X-ray diffraction analysis was performed with a D4 Endeavor with Cu-Kα radiation (λ = 1.54187 Å) & TOPAS software V4.2., Bruker AXS, Germany. In addition to the Rietveld method, also amorphous content and quantitative phase analysis (QPA) was used to evaluate crystalline phases and amorphous contents in the coatings (see [17] for details). High-temperature XRD (HT-XRD) was performed at the PANalytical Empyrean diffractometer in Bragg-Brentano geometry using a Cu Kα anode and an environmental heating chamber HTK1200N (Anton Paar GmbH, Ostfildern, Germany) between room temperature and 1400 °C.

The surface profiles were measured by white light interferometry (cyberTechnologies, CT350T, Eching, Germany).

The EBC systems on oxide/oxide composites were also thermally cycled at 1200°C in a furnace. The heating and cooling rates were +/− 10 K/min, immediately after reaching room temperature the next cycle was started. 4 of these cycles have been made for each coating system. Thermal cycling is seen as a tool to characterize the bonding in the case of the EBC for oxide-based CMCs and hence the efficiency of the surface treatment.

3. Results and Discussion

3.1. APS Coatings on Ox/Ox CMCs

Coatings of Y_2O_3, $Gd_2Zr_2O_7$, $Y_3Al_5O_{12}$ and $Yb_2Si_2O_7$ were applied on the untreated and cleaned substrates. XRD measurements of as-sprayed samples were carried out. The measurements are plotted in Figure 3, signals that cannot be attributed to the desired phase were marked with an asterisk (*). The XRD measurements of Y_2O_3 and $Gd_2Zr_2O_7$ show the diffraction patterns of the corresponding phases, no signals from secondary phases were observed. The coatings appear to have a high crystallinity. In contrast to this, the XRD measurements of $Y_3Al_5O_{12}$ and $Yb_2Si_2O_7$ coatings show humps, which indicate the presence of large amounts of amorphous phases. Due to the low crystallinity not all phases could be identified. The XRD measurements of the $Y_3Al_5O_{12}$ coating shows the diffraction pattern of $Y_3Al_5O_{12}$ and $YAlO_3$. The formation of the alumina depleted phase $YAlO_3$ is attributed to evaporation of alumina during the spraying process, since the vapor pressure of Al_2O_3 is higher than that of Y_2O_3 [24]. The XRD measurement of the $Yb_2Si_2O_7$ coating shows no reflections of the desired phase. Besides the large amorphous hump, the diffraction pattern of Yb_2O_3 was observed. The high amorphous content of the $Y_3Al_5O_{12}$ and $Yb_2Si_2O_7$ coatings might cause crystallization stress in the coating during thermal treatment.

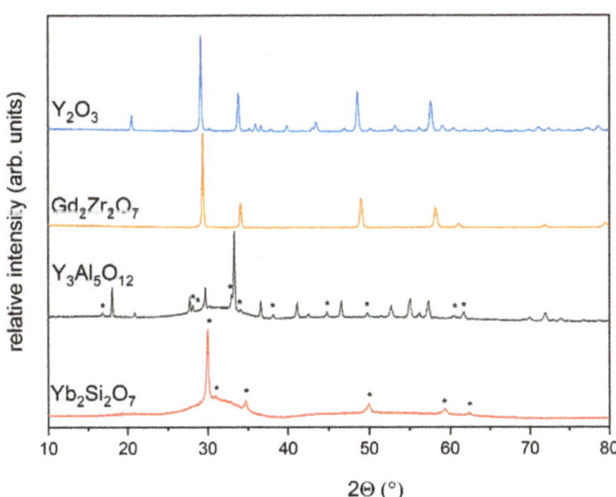

Figure 3. XRD measurements of Y_2O_3, $Gd_2Zr_2O_7$, $Y_3Al_5O_{12}$ and $Yb_2Si_2O_7$ coatings in the as-sprayed state, peaks of secondary phases are marked with *.

The as-sprayed samples were subjected to furnace thermal cycling, the used temperature program consists of four 20 h cycles at 1200 °C. SEM images of cross sections of samples in the as-sprayed state and after thermal cycling are presented in Figure 4. Additionally, higher magnification images of the interface region of the coatings are presented in Figure 5.

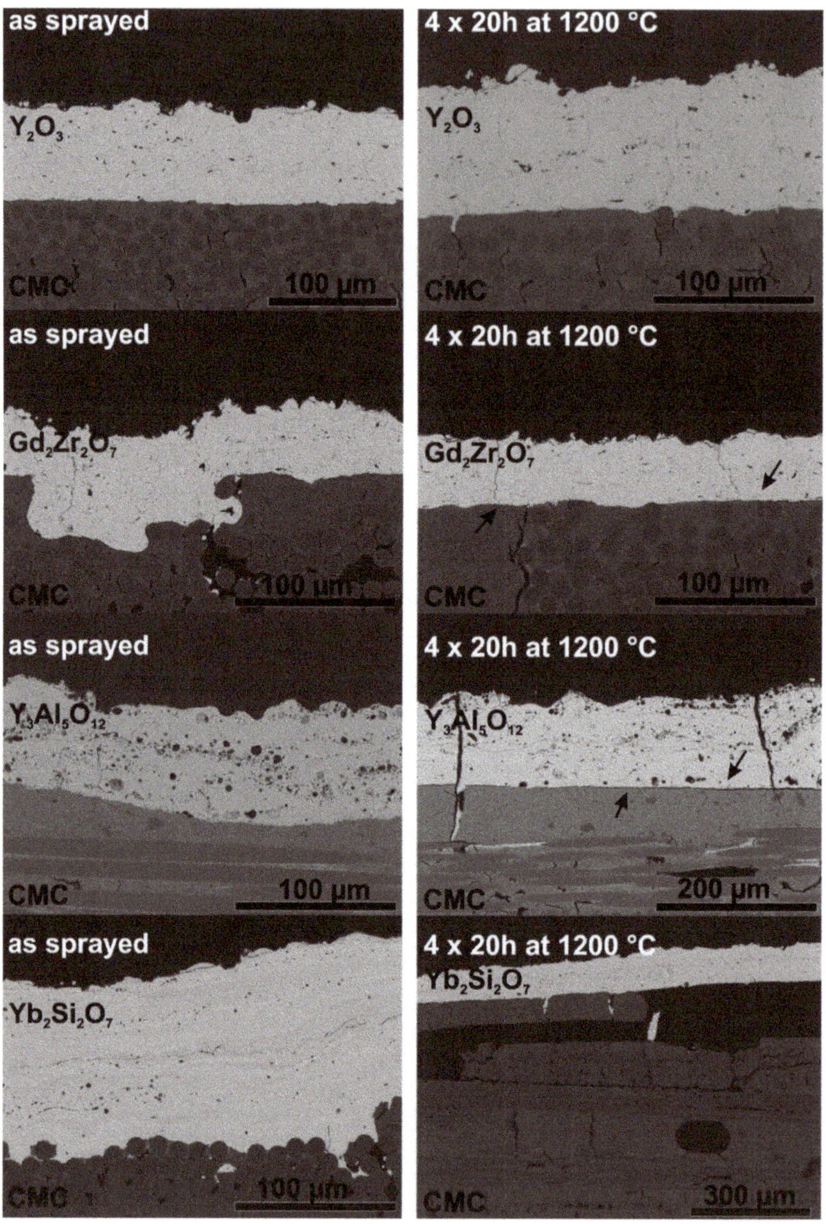

Figure 4. SEM-images of Y_2O_3, $Gd_2Zr_2O_7$, $Y_3Al_5O_{12}$ and $Yb_2Si_2O_7$ APS coatings on Ox/Ox CMCs before (left) and after thermal cycling (right).

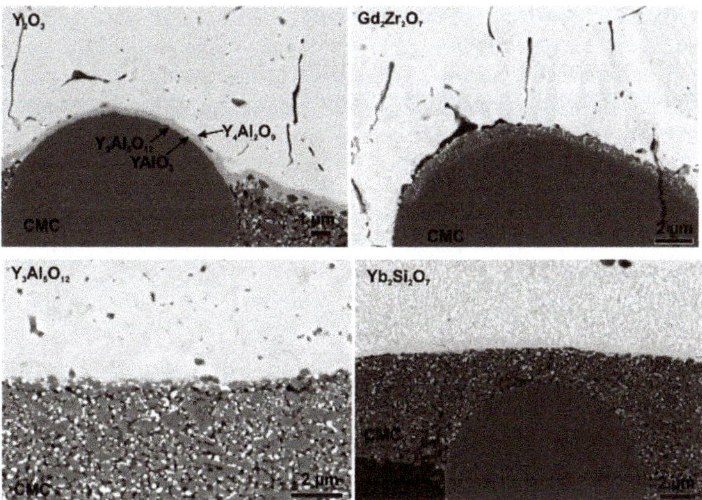

Figure 5. SEM images of the coating substrate interface of Y_2O_3, $Gd_2Zr_2O_7$, $Y_3Al_5O_{12}$ and $Yb_2Si_2O_7$ coatings on Ox/Ox CMCs after thermal treatment.

The obtained Y_2O_3 coatings were dense and fully crystalline. The adhesion of the Y_2O_3 samples seems to be quite good, after heat treatment no signs of cracks or delamination are observed. The good adhesion can be attributed small TEC mismatch and phase stability, resulting in low driving force for delamination. Another reason for the good adhesion is the formation of yttrium-aluminates at the coating substrate interface leading to chemical bonds between coating and substrate [25]. In Figure 5 the formation of several yttrium-aluminates known from the phase diagram [26] can be observed.

The SEM images of the $Gd_2Zr_2O_7$ coatings show dense and homogenous coatings in the as-sprayed state. Figure 5 shows the coating substrate interface in detail. After thermal cycling no sign of a reaction between $Gd_2Zr_2O_7$ and Al_2O_3 was observed. This is in contrast to Lakiza et al. [27] and Leckie et al. [28] The phase diagram of $Gd_2Zr_2O_7$, presented by Lakiza et al. [27], suggests formation of gadolinium-aluminates such as $GdAlO_3$ and $Gd_4Al_2O_9$. The reaction between $Gd_2Zr_2O_7$ and Al_2O_3 to $GdAlO_3$ was observed in a study by Leckie et al. [28]. The absence of this reaction in this study can be explained by bad wetting which causes only small contact areas to the substrate, as a consequence the reaction is inhibited. However, after heat treatment the coating was delaminated at the interface (marked by arrows in Figure 4). This can be explained by the TEC mismatch ($\Delta_{CTE} = 3 \times 10^{-6}$ K^{-1}, see Equations (2) and (3)) and bad wetting of the relative smooth substrate ($Ra - 2.6$ µm).

The SEM images of the $Y_3Al_5O_{12}$ coating confirm the presence of secondary phases within the coating. Furthermore, large round pores can be observed in the $Y_3Al_5O_{12}$ coatings. The high porosity is not favourable for an EBC, as pores and cracks increase the permeability for water vapor. Despite the low TEC mismatch ($\Delta_{CTE} = 3 \times 10^{-6}$ K^{-1}), the $Y_3Al_5O_{12}$ coatings tend to fail during thermal cycling. This can be explained by the high amorphous content of the APS-$Y_3Al_5O_{12}$ coatings. Thermal treatment leads to crystallization and phase transformation of the coating; this causes stress within the material leading to the formation of segmentation and delamination cracks. The formation of large pores in $Y_3Al_5O_{12}$ coatings was also reported by Weyant et al. [29], a dependence between crystallization, substrate temperature and porosity was assumed. According to the phase diagram [30] of Y_2O_3 and Al_2O_3, a reaction between coating and substrate is not expected. The SEM image of the $Y_3Al_5O_{12}$ - Al_2O_3 interface (Figure 5) shows that unlike with Y_2O_3, no reaction occurs between substrate and coating.

The cross-sections show that the disilicate coatings have a lot of pores and consist of several phases, which is, again, not favourable for an EBC. Furthermore, crystallization problems occurred, as described for the $Y_3Al_5O_{12}$ coatings. Due to the high TEC mismatch ($\Delta_{CTE} = 5.2 \times 10^{-6}$ K^{-1}, see Equations (2) and (3)) and the stresses that arise during crystallization, these coatings were delaminated during thermal treatment. According to the phase diagram of Yb_2O_3-SiO_2-Al_2O_3 [31], the formation of mullite ($Al_6Si_2O_{13}$) and ytterbium-aluminates ($Yb_4Al_2O_9$, $Yb_3Al_5O_{12}$) during thermal cycling is possible. However, no reaction between coating and substrate was observed (see Figure 5). The absence of a reaction layer may be attributed to the loose contact between coating and substrate as well as the high TEC mismatch and the resulting premature coating delamination.

Due to the high coating crystallinity, phase purity and coating density, $Gd_2Zr_2O_7$ and Y_2O_3 were chosen for further experiments. Although $Gd_2Zr_2O_7$ has a higher CTE mismatch, it is a desirable top coat due to its excellent CMAS stability. In order to increase the coating adhesion different methods of surface preparation were carried out before spraying.

First, samples of the Al_2O_3/Al_2O_3 CMC were mechanically treated to increase the surface roughness. Some samples of the Al_2O_3/Al_2O_3 CMC were ground, others were grit-blasted. Both methods were able to increase the surface roughness from 2.6 µm to 5.1 µm and 15 µm, respectively. The samples were coated and subsequently furnace cycled for 4 × 20 h at 1200 °C. SEM images of cross-sections of the as-sprayed Y_2O_3-coatings as well as furnace cycled samples are shown in Figure 6. The results show that the mechanical pre-treatment damaged the ceramic substrate to such an extent that even Y_2O_3 coatings failed. Unlike in the study of Gérendas et al. [32], mechanical pre-treatment was not able to achieve sufficient coating adhesion on this CMC.

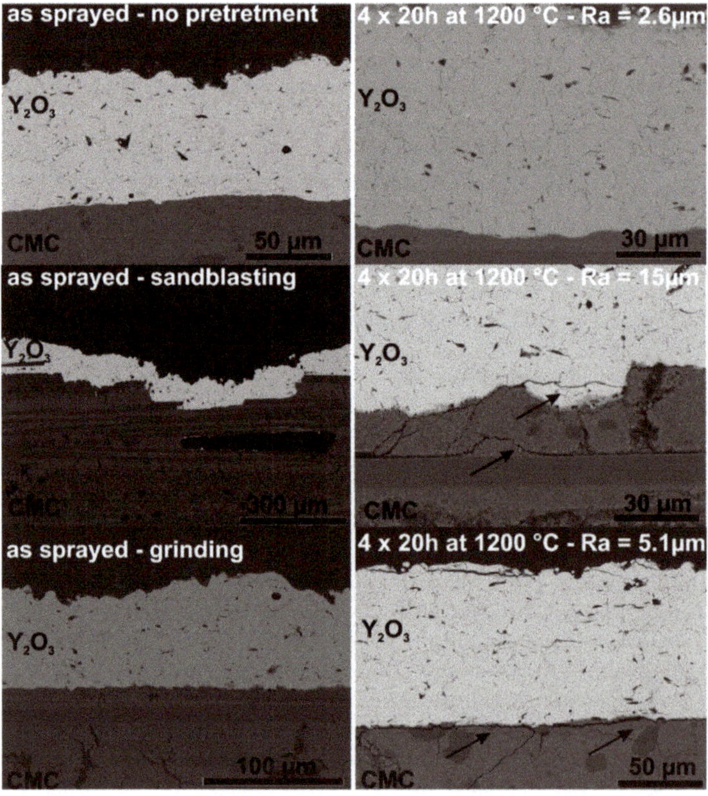

Figure 6. SEM images of APS-Y_2O_3 coatings on Ox/Ox CMCs with and without mechanical pre-treatment before and after thermal cycling.

Laser ablation was used for surface structuring without damaging of the substrates. The positive effect of laser surface structuring of Al_2O_3/Al_2O_3 CMCs on coating adhesion was recently demonstrated by Gatzen et al. [23]. Two different structures were chosen for this study: honeycomb and cauliflower structure. Both structures are illustrated in Figure 7.

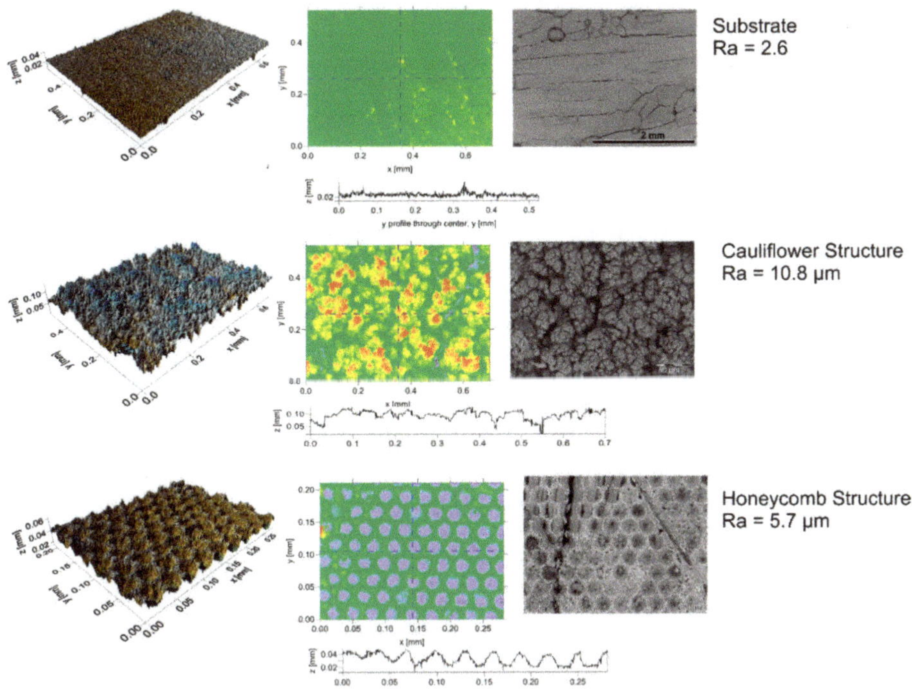

Figure 7. Surface profiles, white light topography and SEM images of ox/ox CMC before (top) and after structuring by laser ablation (middle, bottom).

The untreated substrate has a roughness of about 2.6 μm and a homogenous smooth surface. Few cracks and pores at the surface offer possibilities for clamping. The cauliflower like surface structure comprises an irregular and inhomogeneous surface of re-solidified alumina. An average roughness of 10.8 μm was achieved with this pattern. The honeycomb structure shows well- defined holes close to each other. A roughness of about 5.7 μm was measured for this pattern. Both surface structures offer more possibilities for clamping, therefore, an increase in lifetime is expected especially for $Gd_2Zr_2O_7$ coatings.

After structuring the samples were coated with Y_2O_3 and $Gd_2Zr_2O_7$. SEM images of the coated and thermally cycled samples are shown in Figure 8. Both coatings could infiltrate the voids in between the substrate structure. The surface structures, especially the cauliflower structure, allow interlocking of the coating. After thermal cycling no cracks or delamination occurred in the Y_2O_3 coating-substrate-system. This shows that in contrast to grit-blasting and grinding, laser structuring of the CMC does not weaken the ceramic matrix. Furthermore, after thermal cycling of the laser-structured $Gd_2Zr_2O_7$ coated samples, the coating-substrate interface shows no delamination cracks. This is a significant improve, compared to the coating on untreated substrates. Although the energy release rate is high in these coatings with high mismatch (see Equations (2) and (3)), the laser structuring of the samples proofed to be beneficial for the coating adhesion (e.g., the critical energy release).

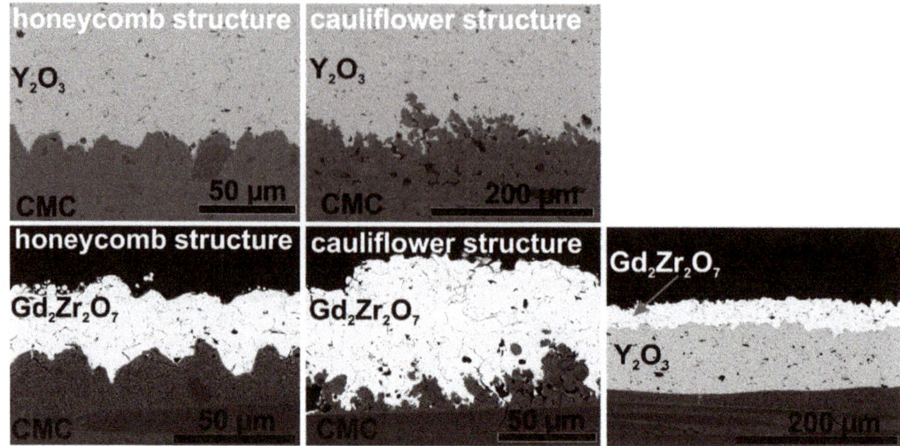

Figure 8. SEM images of APS Y_2O_3 and $Gd_2Zr_2O_7$ coatings on laser structured Ox/Ox CMC substrates after thermal cycling (4 × 20 h at 1200 °C).

In addition, a third method of adhesion improvement was tested for $Gd_2Zr_2O_7$: the usage of an Y_2O_3-bondcoat. For this, the untreated substrate was first coated with Y_2O_3 and then coated with $Gd_2Zr_2O_7$. This double layer coating system is referring to the double layered TBCs as published by Vaßen et al. [33]. The Y_2O_3-bondcoat helps to buffer to some extent especially at edges the CTE mismatch between the CMC and $Gd_2Zr_2O_7$. Furthermore, the Y_2O_3 coating is known to be good adherent on Alumina CMCs [25] and offer a rough surface for the $Gd_2Zr_2O_7$ coating. Using Y_2O_3 as an interlayer between substrate and $Gd_2Zr_2O_7$ also proofed to increase the coating adhesion as well.

3.2. Development of APS Y_2SiO_5 coatings

According to Figure 1 Y_2SiO_5 has an excellent water vapor resistance and is therefore considered as appropriate EBC material [34,35]. In this investigation only rather hot APS spraying conditions are investigated for this material. The density of all the coatings manufactured by the investigated conditions (see Table 2) has been rather high, more interesting is the phase content which shows remarkable differences (see Figure 9). Compared to the He as secondary gas, hydrogen increased the plasma enthalpy which leads to higher substrate temperatures (750°C) and increases the amount of crystallization. The comparable amount of SiO_2 in all coatings might indicate no significant differences in the silica loss during spraying (although the amorphous content is not considered).

Due to the highest crystallinity, the condition D with hydrogen and long stand-off distance (condition D) was used for further studies and coatings on C/SiC substrates were prepared. In Figure 10 SEM micrographs of the coatings in the as-sprayed condition and after annealing for 12 h at 1350 °C. First of all, the coating is also after the heat-treatment well-adherent. This indicates that the TEC mismatch (see Equations (2) and (3)) between coating and substrate seems to be not extremely detrimental (although the material would fit even better to Ox/Ox CMC substrates). In addition, also the phase transformation at 850 °C and the strongly anisotropic expansion of Y_2SiO_5 [36] obviously does not significantly affect coating integrity.

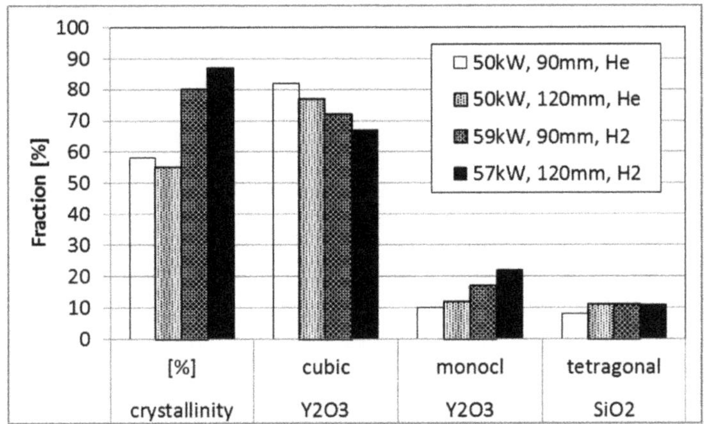

Figure 9. Phase evaluation of 4 Y_2SiO_5 coatings (caption indicate power, stand-off distance and secondary gas).

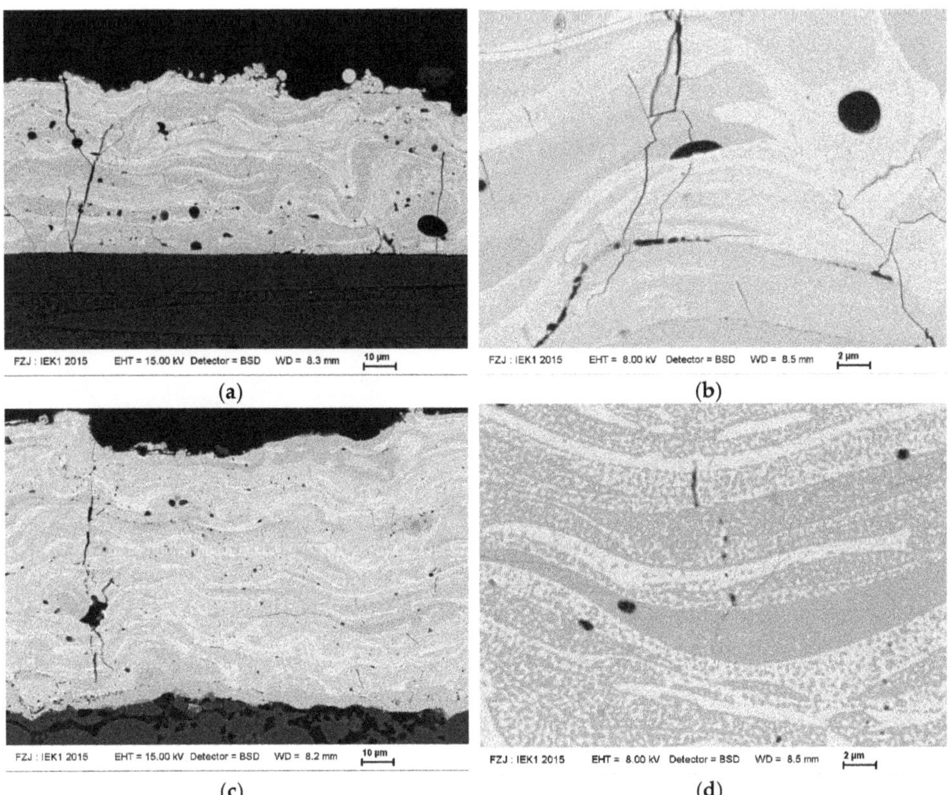

Figure 10. SEM micrographs of plasma-sprayed Y_2SiO_5 coatings (condition D) in the as-sprayed (**a,b**) and annealed (**c,d**, 1350°C, 12h) condition.

In addition, the size of the segmentation and micro cracks is largely reduced due to sintering, some segmentation cracks are even no longer penetrating through the whole coating. Also the phase distribution changed significantly. While in the as-sprayed condition rather large regions with low and high silicon content appear (stemming from the impinging particles which have lost silicon during spraying mainly at their surface), the annealed coating shows much finer phases consisting of cubic Y_2O_3 (bright) and monoclinic Y_2SiO_5 (darker). The evolution of phases is shown in Figure 11 measured by XRD and using Rietveld refinement. Obviously, a short-term annealing at 1000 °C is not sufficient to form the equilibrium phases. As the coating is rather dense, the Y_2O_3 as a second phase has a reasonable corrosion resistance (Figure 1) and the microstructure probably will show some particulate toughening effects, the found coating is expected to perform good as EBC.

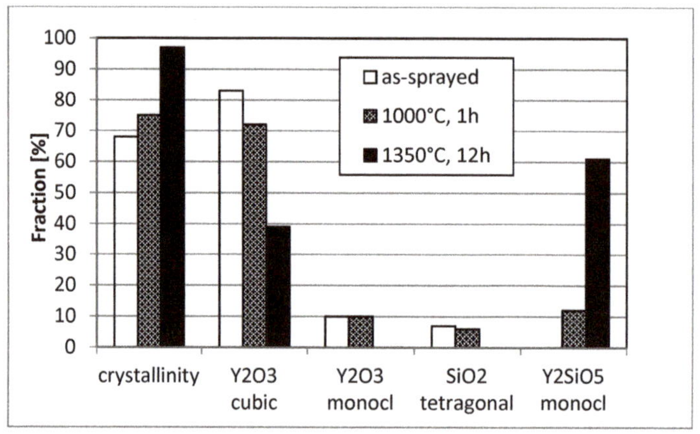

Figure 11. Phase evolution in APS Y_2SiO_5 coatings (condition D) after heat-treatment.

3.3. $Yb_2Si_2O_7$ Coatings Manufactured by Different Thermal Spray Techniques for SiC/SiC Substrates

Figure 12 shows the phase composition of APS, HVOF and VLPPS coatings deposited from different particle size fractions of the same $Yb_2Si_2O_7$ feedstock. The main differences between the different methods are melting degree of the sprayed particles and deposition temperatures which define the crystallinity of the coatings as well as the degree of Si evaporation during spraying.

In APS and VLPPS, as a result of the high heat transfer from plasma to the particles, particles melt, impact on the substrate in the molten state and rapidly solidify on the substrate. High plasma powers were selected in the deposition of EBCs with APS (57 kW) and VLPPS (90 kW) as a dense microstructure consists of well-flattened particles with good interfacial contact is aimed. While high plasma powers ensure melting of particles in the plume, it also leads to a significant amount of Si-evaporation from $Yb_2Si_2O_7$ during spraying. As a result of that, Si-depleted secondary phases such as Yb_2SiO_5 and Yb_2O_3 are found in the as-sprayed coatings as shown in the Figure 12a. Furthermore, due to quenching of the molten particles on the substrate, that is at nearly room temperature if not pre-heated by plasma plume, crystallization of the glass forming silicates is suppressed and the amorphous phase is formed. The APS experiments carried out in order to increase the SiC substrate temperature by heating with plasma prior to deposition revealed that for a substrate size of $50 \times 50mm^2$, it is possible to increase the substrate temperatures up to about 800–900 °C however the temperature rapidly goes down to 500–550 °C till the deposition starts. As shown in Figure 12a, this deposition temperature range was found to be not sufficient to activate and complete the crystallization of the coating. According to HT-XRD analysis of amorphous plasma-sprayed $Yb_2Si_2O_7$ particles, which were collected in water and dried subsequently at 70 °C, the crystallization temperature of the material was found to be above 1000 °C, which explains the amorphous structure of the coating deposited at about 550 °C. Aiming at a higher substrate temperature as well as slower cooling rates in order to provide higher energy and longer time for the atoms to rearrange into the crystalline state, VLPPS experiments were conducted in the controlled atmosphere chamber (2 mbar). It was found that it is possible to reach substrate temperatures higher than 1000 °C (Figure 13) owing to higher plasma power of the O3CP torch as well as to retain it till the deposition starts due to reduced heat transfer under vacuum. To prolong the high-temperature phase after deposition for some minutes, the coating was kept to be heated using the plasma plume and by this procedure, highly crystalline coatings could be produced as shown in Figure 12a. Nevertheless, it should be mentioned that Si-evaporation remains a problem as more than 10 wt.% Yb_2SiO_5 was detected in the VLPPS coatings.

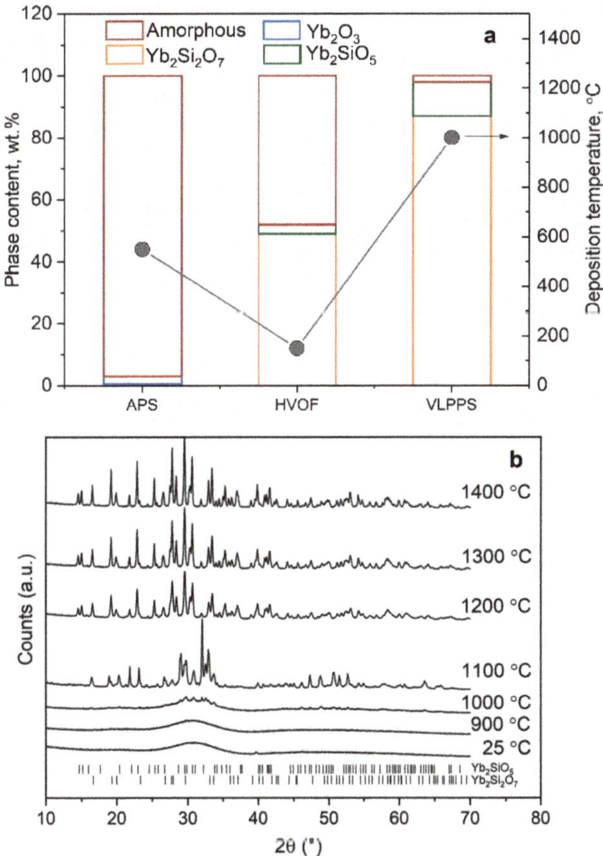

Figure 12. (**a**) Quantitative phase composition of the as-deposited coatings from $Yb_2Si_2O_7$ feedstock using different thermal spray methods, (**b**) HT-XRD diffractograms of atmospheric plasma sprayed $Yb_2Si_2O_7$ feedstock.

The HVOF process yields lower flame temperatures in comparison with the plasma spray processes as the flame is generated by combustion. As a result of that, particle temperatures as well as the deposition temperature are lower. Nonetheless, using the HVOF process, $Yb_2Si_2O_7$ coatings with higher crystallinity in comparison with the APS could be manufactured because of the deposition of un-melted or partially molten particles. This was avoided in the plasma spray processes as un-melted particles increase the porosity levels in the coatings due to imperfect contact regions as well as the porous morphology of the particle itself. In the HVOF process, however, not the porosity stemming from the particle morphology, but the porosity caused by the bad contact can be minimized thanks to very high flame velocity and, thus, the high momentum transfer to the particles in the flame. If the brittle oxide particles are completely un-melted, they break upon high-velocity impact on the substrate, however, if only their core is un-melted and the outer surface is molten, they can adhere on the substrate or on the previously deposited layers. To reach such particle conditions, sensitive process optimizations are required in terms of fuel–oxygen stoichiometry, total gas flow, powder feed rate, stand-off distance, and particle size distribution. Details of these investigations can be found elsewhere [18] and HVOF microstructure will be further discussed in the following. Consequently, these coatings own higher crystallinity regardless of the lower deposition temperature in the process as un-melted part of the particles remain crystalline. An added benefit of low flame temperature is diminished Si-evaporation from the particles during spraying. It can be seen from Figure 12a that the detected Yb_2SiO_5 content from about 50 wt.% crystalline HVOF coating is about 5 wt.% which fits well to the feedstock composition although of course the large amorphous content has to be considered.

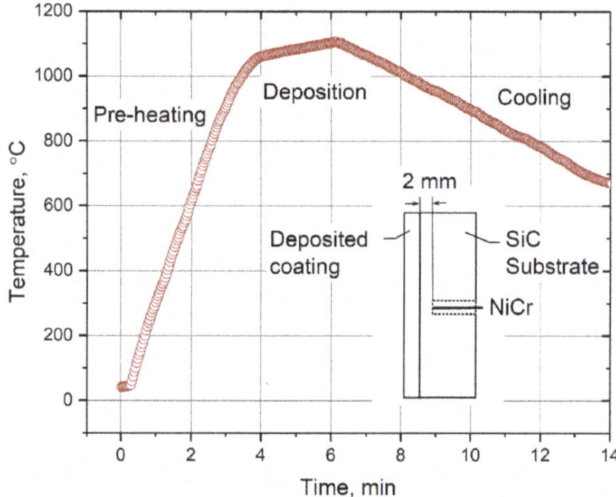

Figure 13. Recorded temperature measurements using a thermocouple during VLPPS deposition of $Yb_2Si_2O_7$ on monolithic SiC substrate.

The microstructure of the coatings sprayed from $Yb_2Si_2O_7$ feedstock with APS, HVOF and VLPPS processes can be seen in Figure 14. The two APS coatings (a,b) that were deposited at different temperatures (Figure 14a is the "standard" APS sample and b is deposited at a higher temperature, further details are discussed below) have vertical cracks running through the thickness of the coatings which is associated with the higher thermal stresses in these coatings due to the significant amount of Yb_2SiO_5 content (about 30 wt.% in these coatings determined after crystallization heat treatment). Because thermal expansion coefficient of Yb_2SiO_5 (7.5×10^{-6} °C^{-1}) is larger than that of the $Yb_2Si_2O_7$ (4.7×10^{-6} °C^{-1}) as well as the SiC/SiC substrate (5.1×10^{-6} °C^{-1}) [37]. This leads to greater tensile thermal stresses in the oxide coating after cooling, along with the tensile quenching stresses, which induce the vertical cracking. The high amorphous content in the APS layer would even further increase the CTE mismatch, however simultaneously due to a reduced Young's modulus of the amorphous state, the effect on the stress level might be limited. Therefore, as cracks are not desired in the EBC microstructure, it is of crucial importance to minimize Si- evaporation during spraying.

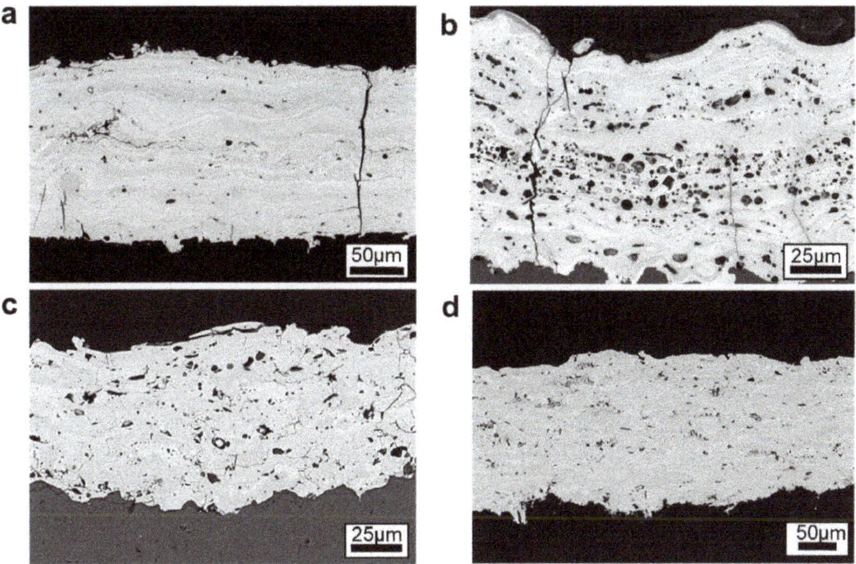

Figure 14. Microstructure of the coatings sprayed from $Yb_2Si_2O_7$ feedstock using (**a,b**) APS, (**c**) HVOF and (**d**) VLPPS process. Note that both APS coatings (**a,b**) were sprayed with 520 A current and 90 mm spray distance but deposition temperatures were about 550 °C and 900 °C at a and b, respectively (see text).

Both APS coatings (Figure 14a,b) were sprayed at similar conditions but sample size in (b) was very small ($3 \times 3.8 \times 36$ mm^3) and as a result of that a maximum deposition temperature of about 900 °C could be reached for this particular sample. Therefore, in contrast to standard size APS sample that was analysed in the preceding section and shown in Figure 14a, the APS coating shown in Figure 14b is highly crystalline (76 wt.%). The porosity content of the two APS coatings are also dissimilar, i.e., the coating (a) is significantly denser than the coating (b). This difference between the microstructures was associated with the higher crystallinity in the coating (b). Seemingly, heterogeneous nucleation takes place at the splat boundaries during cooling while the centre of the splats remains amorphous. As density increases in the crystallizing zone, the volume is reduced which induces elastic tensile stresses within the amorphous region. Pores within the splat, therefore, nucleate when these stresses are large enough. Other relevant theories to the formation of pores are discussed elsewhere [20].

No vertical cracking was observed in the fairly dense HVOF layer (Figure 14c), presumably due to lower Yb_2SiO_5 content and hence reduced thermal stresses in the process as a result of the lower particle and deposition temperatures, respectively. However, isothermal thermal cycling experiments revealed that the adhesion of the HVOF oxide layer on the Si bond coat is poor because the HVOF layer partially delaminated only after few cycles [18]. The short lifetime of the layer was attributed to the presence of un-melted particles at the interface wherein the cracks can easily propagate. The un-melted particles can be avoided for instance by decreasing the particle size of the feedstock but that means higher amorphous content and Yb_2SiO_5 phase in the layer at the same time. Figure 14d shows the microstructure of a VLPPS layer that is free of vertical cracks, dense and also crystalline in the as-sprayed state, which makes it a very promising method developed for EBC manufacturing in Jülich. High deposition temperatures (min. 1000 °C) and moderate cooling rates (approx. 55 K/min) in the process as shown in Figure 13 evidently helps for crystallization as well as for stress relaxation in the layer. Further investigations are ongoing in this direction to understand the effect of deposition temperature and cooling rates on crystallization kinetics and related mechanical properties in the layer, as well as the stress state in the EBC system.

While a variety of coating morphologies can be obtained with thermal spray processes, characteristics of the coatings, typically the size of microstructural features, is controlled by the used feedstock [38]. For example, the size of intersplat pores is directly depending on the size and shape of powdery feedstocks. Since a well flowable powder in the size range of about 10–100 μm is usually applied for conventional plasma spraying, there is a minimum size for application of powdery feedstock. Suspension plasma spray would be one of alternative processes for conventional ones in this context. Suspension plasma spraying is one of the rather new thermal spray techniques, which has a liquid feed stock and relatively higher plasma power compared with conventional ones. Especially, SPS could supply a great diversity in microstructural features from columnar and porous to bulk-like dense ones [39,40]. Main processing parameters, such as gun power, spraying distance, roughness, and particle size distribution in suspension, would play an important role in controlling the microstructures fabricated by SPS.

Figure 15 shows the cross-sectional microstructures of $Yb_2Si_2O_7$ coatings fabricated by using the SPS technique with different spraying distances of 50 and 70 mm. As seen especially in the higher magnification images, bulk-like dense microstructure can be obtained within this work regarding gun power, spraying distance and particle size distribution in suspension feedstock. Vertical cracks in the coating microstructure were inevitable because of tensile stress from rapid cooling after spraying as can be seen in Figure 15c.

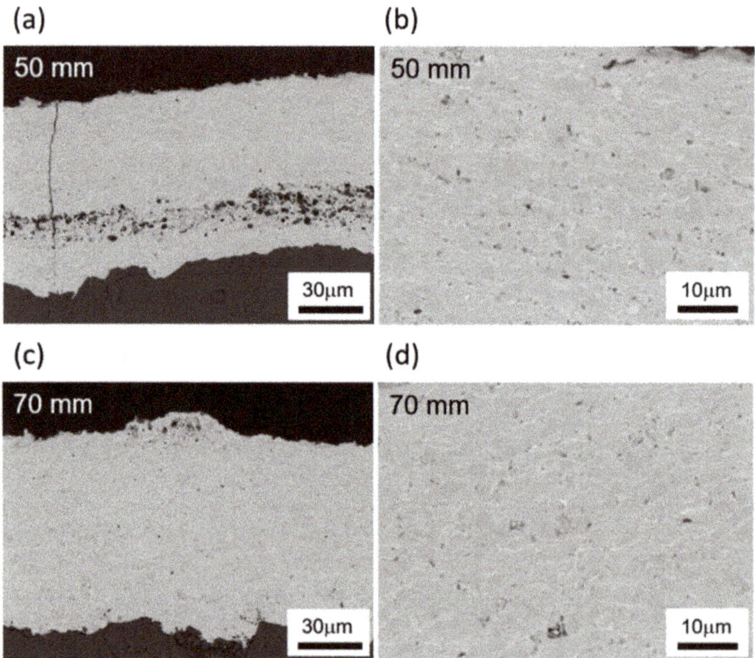

Figure 15. Microstructure of the coatings sprayed from $Yb_2Si_2O_7$ suspension by using SPS with a gun distance of (**a**,**b**) 50 mm and (**c**,**d**) 70 mm.

Figure 16 reveals how the stand-off distance influences the substrate temperature and by that e.g., the degree of crystallization. Shorter stand-off distances give higher temperatures and by that also better splat bonding and reduced relaxation, which give a higher segmentation crack density. Figure 17 shows the X-ray diffraction patterns of $Yb_2Si_2O_7$ coatings fabricated by using the SPS technique with different spraying distance. Considerable amounts of the amorphous phase were observed from EBC with long spraying distance (90 mm). In the case of short spraying distance (50 mm), the degree of crystallinity was around 80%, which was calculated from the areal intensity ratio between crystalline peaks and amorphous humps of XRD.

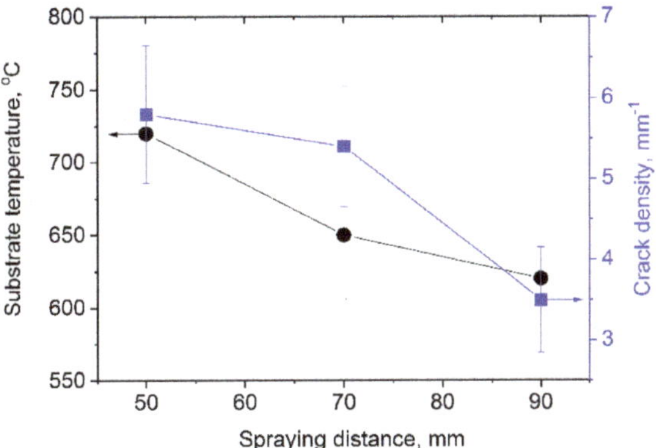

Figure 16. Substrate temperature and crack density as function of the spraying distance.

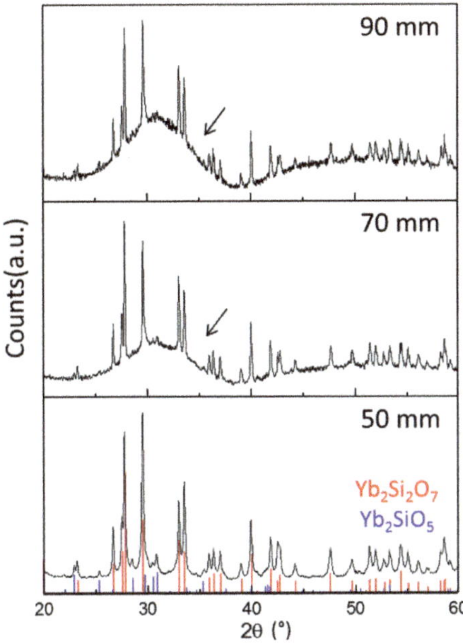

Figure 17. X-ray diffraction patterns from YDS EBCs fabricated by using suspension plasma spraying with spraying distance of 50, 70, or 90 mm. Arrows indicate amorphous baselines in XRD.

4. Conclusions

This study summarizes insights into the development of EBCs for both oxide- and non-oxide-based CMCs. Results of coating experiments with different ceramic powders on an Al_2O_3/Al_2O_3 CMC were presented; the samples were furnace cycled to test the high temperature behaviour. Coatings of Y_2O_3 and $Gd_2Zr_2O_7$ showed promising results, as they were crystalline and dense in the as-sprayed state. Y_2O_3 coatings showed excellent adhesion due to the formation of chemical bonds between coating and substrate. In contrast to this, $Gd_2Zr_2O_7$ coatings tend to fail during cycling because of bad contact between coating and substrate. The coating adhesion could be significantly increased by laser structuring of the CMC before coating.

Y_2SiO_5 coatings could be prepared with rather high crystallinity (nearly 70%). Heat treatment led to the formation of a fine-grained microstructure with the major phase Y_2SiO_5 and a considerable amount of Y_2O_3 without delamination after cooling.

$Yb_2Si_2O_7$ coatings have been produced by different thermal spray process, namely, APS, HVOF, VLPPS, and SPS. APS can deliver dense coatings with a high degree of amorphous phase content. HVOF gives rather crystalline coatings with acceptable porosity levels, VLPPS gives even better coatings. Furthermore, SPS can give a high degree of crystallinity, however segmentation cracks are difficult to avoid.

A summary of the outcome reflecting the actual situation as seen in our institute is given in Table 4. So, our message is that also other thermal spray methods than APS can be used to obtain promising EBCs. Certainly, further improvements of processes can change the ranking in this evaluation.

Table 4. Comparison of the used thermal spray processes for the deposition of EBCs (o reflects average, - (– very) bad, + (++ very) good results with respect to the criteria).

Spray Method	Crystallinity	Density	Cracks	Costs	Overall
APS	0	0	-	+	0
SPS	+	+	-	0	0+
VLPPS	++	++	+	–	+
HVOF	+	+	+	0	+

Author Contributions: R.V.: writing—original draft preparation, supervision, review and editing; E.B.: investigation, writing—original draft preparation; C.G.: investigation, writing—original draft preparation; S.K.: investigation, writing—original draft preparation; D.E.M.: supervision, review and editing; O.G.: supervision, review and editing.

Funding: This research received no external funding.

Acknowledgments: The authors acknowledge the contributions of the following colleagues in our institute: Ralf Laufs, Frank Kurze and Mr. Karl-Heinz Rauwald for the invaluable assistance during plasma spraying and Georg Mauer for valuable discussions of thermal spray results. We also would like to thank Doris Sebold for SEM analysis and Yoo Jung Sohn for the extended XRD analysis. We also appreciate the support of Sigrid Schwartz-Lückge and Mark Kappertz in sample preparation and characterization.

Conflicts of Interest: The authors declare no conflict of interest.

References

1. Bansal, N.P. *Handbook of Ceramic Composites*; Kluwer Academic Publishers: Boston, MA, USA; Dordrecht, The Netherlands; London, UK, 2005.
2. Steibel, J. Ceramic matrix composites taking flight at GE Aviation. *Am. Ceram. Soc. Bull.* **2019**, *98*, 30–33.
3. Gerendás, M.; Wilhelmi, C.; Machry, T.; Knoche, R.; Werth, E.; Behrendt, T.; Koch, D.; Hofmann, S.; Göring, J.; Tushtev, K.; et al. Dwvelopment and Validation of Oxide/Oxide CMC Combustors within the Hipoc Program. In Proceedings of the ASME Turbo Expo 2013, San Antonio, TX, USA, 3–7 June 2013.
4. Opila, E.J.; Hann Jr, R.E. Paralinear Oxidation of CVD SiC in Water Vapor. *J. Am. Ceram. Soc.* **1997**, *80*, 197–205. [CrossRef]
5. Herrmann, M.; Klemm, H. Corrosion of Ceramic Materials. In *Comprehensive Hard Materials*; Sarin, V.K., Ed.; Elsevier: Oxford, UK, 2014.
6. Zhu, D.; Fox, D.S.; Bansal, N.P.; Miller, R.A. *Advanced Oxide Material Systems for 1650 °C Thermal/Environmental Barrier Coating Application*; NASA Technical Report 20050199717; NASA Glenn Research Center: Cleveland, OH, USA, 2004.
7. Lehmann, H.; Pitzer, D.; Pracht, G.; Vaßen, R.; Stöver, D. Thermal Conductivity and Thermal Expansion Coefficients of the Lanthanum Rare-Earth-Element Zirconate System. *J. Am. Ceram. Soc.* **2003**, *86*, 1338. [CrossRef]

8. Fritsch, M.; Klemm, H.; Herrmann, M.; Michaelis, A.; Schenk, B. The water vapour hot gas corrosion of ceramic materials. *Ceram. Forum Int.* **2010**, *87*, 11–12.

9. Fritsch, M.; Klemm, H.; Herrmann, M.; Schenk, B. Corrosion of selected ceramic materials in hot gas environment. *J. Eur. Ceram. Soc.* **2006**, *26*, 3557–3565. [CrossRef]

10. Fritsch, M. *Heißgaskorrosion Keramischer Werkstoffe in H2O-Haltigen Rauchgasatmosphären, Fraunhofer IRB Verlag;* TU Dresden: Dresden, Germany, 2007.

11. Fauchais, P.L.; Heberlein, J.V.; Boulos, M.I. *Thermal Spray Fundamentals;* Springer: New York, NY, USA, 2014.

12. Li, C.-J.; Yang, G.-J.; Li, C.-X. Development of Particle Interface Bonding in Thermal Spray Coatings: A Review. *J. Therm. Spray Technol.* **2013**, *22*, 192–206. [CrossRef]

13. Vaßen, R.; Hathiramani, D.; Mertens, J.; Haanappel, V.; Vinke, I. Manufacturing of high performance solid oxide fuel cells (SOFCs) with atmospheric plasma spraying (APS). *Surf. Coat. Technol.* **2007**, *202*, 499–508. [CrossRef]

14. Guo, H.B.; Vaßen, R.; Stöver, D. Atmospheric plasma sprayed thick thermal barrier coatings with high segmentation crack density. *Surf. Coat. Technol.* **2004**, *186*, 353–363. [CrossRef]

15. Richards, B.T.; Wadley, H.N.G. Plasma spray deposition of tri-layer environmental barrier coatings. *J. Eur. Ceram. Soc.* **2014**, *34*, 3069–3083. [CrossRef]

16. Marcano, D.; Mauer, G.; Sohn, Y.J.; Vassen, R.; Garcia-Fayos, J.; Serra, J.M. Controlling the stress state of $La_{1-x}Sr_xCo_yFe_{1-y}O_{3-\delta}$ oxygen transport membranes on porous metallic supports deposited by plasma spray–physical vapor process. *J. Membr. Sci.* **2016**, *503*, 1–7. [CrossRef]

17. Bakan, E.; Marcano, D.; Zhou, D.; Sohn, Y.J.; Mauer, G.; Vaßen, R. $Yb_2Si_2O_7$ Environmental Barrier Coatings Deposited by Various Thermal Spray Techniques: A Preliminary Comparative Study. *J. Therm. Spray Technol.* **2017**, *26*, 1011–1024. [CrossRef]

18. Bakan, E.; Mauer, G.; Koch, D.; Vassen, R.; Sohn, Y.J. Application of High-Velocity Oxygen-Fuel (HVOF) Spraying to Fabrication of $Yb_2Si_2O_7$ Environmental Barrier Coatings. *Coatings* **2017**, *7*, 55. [CrossRef]

19. Gatzen, C.; Mack, D.E.; Guillon, O.; Vaßen, R. Water vapor corrosion test using supersonic gas velocities. *J. Am. Ceram. Soc.* **2019**, *102*, 6850–6862. [CrossRef]

20. Bakan, E.; Sohn, Y.J.; Kunz, W.; Klemm, H.; Vaßen, R. Effect of processing on high-velocity water vapor recession behavior of Yb-silicate environmental barrier coatings. *J. Eur. Ceram. Soc.* **2019**, *39*, 1507–1513. [CrossRef]

21. Mainzer, B.; Frieß; Martin; Jemmali, R.; Koch, D. Development of polyvinylsilazane-derived ceramic matrix composites based on Tyranno SA3 fibers. *J. Ceram. Soc. Jpn.* **2016**, *124*, 1035–1041. [CrossRef]

22. Rüdinger, W.P.A. Die Entwicklung oxidkeramischer Faserverbundwerkstoffe am Fraunhofer ISC/Zentrum HTL in Zusammenarbeit mit W.E.C. Pritzkow Spezialkeramik. *Keramische Zeitschrift* **2013**, *03*, 166–169.

23. Gatzen, C.; Mack, D.E.; Guillon, O.; Vaßen, R. Surface roughening of Al_2O_3/Al_2O_3-ceramic matrix composites by nanosecond laser ablation prior to thermal spraying. *J. Laser Appl.* **2019**, *31*, 022018. [CrossRef]

24. Schulz, U.; Saruhan, B.; Fritscher, K.; Leyens, C. Review on Advanced EB-PVD Ceramic Topcoats for TBC Applications. *Int. J. Appl. Ceram. Tec.* **2004**, *1*, 302–315. [CrossRef]

25. Mechnich, P.; Braue, W. Air plasma-sprayed Y_2O_3 coatings for Al_2O_3/Al_2O_3 ceramic matrix composites. *J. Eur. Ceram. Soc.* **2013**, *33*, 2645–2653. [CrossRef]

26. Fabrichnaya, O.; Seifert, H.J.; Weiland, R.; Ludwig, T.; Aldinger, F.A. Navrotsky, Phase Equilibria and Thermodynamics in the Y_2O_3-Al_2O_3-SiO_2 System. *Zeitschrift für Metallkunde* **2001**, *92*, 1083–1097.

27. Lakiza, S.; Fabrichnaya, O.; Wang, C.; Zinkevich, M.; Aldinger, F. Phase diagram of the ZrO_2–Gd_2O_3–Al_2O_3 system. *J. Eur. Ceram. Soc.* **2006**, *26*, 233–246. [CrossRef]

28. Leckie, R.M.; Krämer, S.; Rühle, M.; Levi, C.G. Thermochemical compatibility between alumina and ZrO_2–$GdO_{3/2}$ thermal barrier coatings. *Acta Mater.* **2005**, *53*, 3281–3292. [CrossRef]

29. Weyant, C.M.; Faber, K.T. Processing–microstructure relationships for plasma-sprayed yttrium aluminum garnet. *Surf. Coat. Technol.* **2008**, *202*, 6081–6089. [CrossRef]

30. Fabrichnaya, O.; Seifert, H.J.; Ludwig, T.; Aldinger, F.; Navrotsky, A. The assessment of thermodynamic parameters in the Al_2O_3-Y_2O_3 system and phase relations in the Y-Al-O system. *Scand. J. Metall.* **2001**, *30*, 175–183. [CrossRef]

31. Murakami, Y.; Yamamoto, H. Phase Equilibria and Properties of Glasses in the Al_2O_3-Yb_2O_3-SiO_2 System. *J. Ceram. Soc. Jpn.* **1993**, *101*, 1101–1106. [CrossRef]

32. Gerendás, M.; Cadoret, Y.; Wilhelmi, C.; Machry, T.; Knoche, R.; Behrendt, T.; Aumeier, T.; Denis, S.; Göring, J.; Koch, D.; et al. Improvement of OxideOxide CMC and Development of Combustor and Turbine Components in the HiPOC Program. *ASME Turbo Expo* **2011**, *1*, 477–490.
33. Vaßen, R.; Traeger, F.; Stöver, D. New Thermal Barrier Coatings Based on Pyrochlore/YSZ Double-Layer Systems. *Int. J. Appl. Ceram. Technol.* **2004**, *1*, 351–361. [CrossRef]
34. Garcı́a, P.M.E.; Osendi, M.I. The Prospect of Y_2SiO_5-Based Materials as Protective Layer in Environmental Barrier Coatings. *J. Therm. Spray Technol.* **2013**, *22*, 680–689. [CrossRef]
35. Zhang, J.P.; Fu, Q.G.; Zhuang, L.; Li, H.J.; Sun, C. Preparation and Ablation Properties of Y_2SiO_5 Coating for SiC-Coated C/C Composites by Supersonic Plasma Spraying. *J. Therm. Spray Technol.* **2015**, *24*, 994–1001. [CrossRef]
36. Nowok, J.W.; Kay, J.P.; Kulas, R.J. Thermal expansion and high-temperature phase transformation of the yttrium silicate Y_2SiO_5. *J. Mater. Res.* **2001**, *16*, 2251–2255. [CrossRef]
37. Lee, K.N. Enviromental Barrier Coatings For SiC/SiC. In *Ceramic Matrix Composites: Materials, Modeling and Technology*; Bansal, N.P., Lamon, J., Eds.; The American Ceramic Society: Hoboken, NJ, USA, 2015.
38. Vaßen, R.; Kaßner, H.; Mauer, G.; Stöver, D. Suspension Plasma Spraying: Process Characteristics and Applications. *J. Therm. Spray Technol.* **2010**, *19*, 219–225. [CrossRef]
39. Zhou, D.; Guillon, O.; Vaßen, R. Development of YSZ Thermal Barrier Coatings Using Axial Suspension Plasma Spraying. *Coatings* **2017**, *7*, 120–136. [CrossRef]
40. Kim, S.-J.; Lee, J.-K.; Oh, Y.-S.; Kim, S.; Lee, S.-M. Effect of Processing Parameters and Powder Size on Microstructures and Mechanical Properties of Y_2O_3 Coatings Fabricated by Suspension Plasma Spray. *J. Korean Ceram. Soc.* **2015**, *52*, 395–402. [CrossRef]

Article

YAlO$_3$—A Novel Environmental Barrier Coating for Al$_2$O$_3$/Al$_2$O$_3$–Ceramic Matrix Composites

Caren Gatzen *, Daniel Emil Mack, Olivier Guillon and Robert Vaßen

Forschungszentrum Jülich GmbH, Institute of Energy and Climate Research, Materials Synthesis and Processing (IEK-1), 52425 Jülich, Germany
* Correspondence: c.gatzen@fz-juelich.de

Received: 30 August 2019; Accepted: 23 September 2019; Published: 25 September 2019

Abstract: Ceramic matrix composites (CMCs) are promising materials for high-temperature applications. Environmental barrier coatings (EBCs) are needed to protect the components against water vapor attack. A new potential EBC material, YAlO$_3$, was studied in this paper. Different plasma-spraying techniques were used for the production of coatings on an alumina-based CMC, such as atmospheric plasma spraying (APS) and very low pressure plasma spraying (VLPPS). No bond coats or surface treatments were applied. The performance was tested by pull–adhesion tests, burner rig tests, and calcium-magnesium-aluminum-silicate (CMAS) corrosion tests. The samples were subsequently analyzed by means of X-ray diffraction, scanning electron microscopy, and energy-dispersive X-ray spectroscopy. Special attention was paid to the interaction at the interface between coating and substrate. The results show that fully crystalline and good adherent YAlO$_3$ coatings can be produced without further substrate preparation such as surface pretreatment or bond coat application. The formation of a thin reaction layer between coating and substrate seems to promote adhesion.

Keywords: atmospheric plasma spraying (APS); very low pressure plasma spraying (VLPPS); environmental barrier coating (EBC); YAlO$_3$; yttrium aluminum perovskite (YAP); ceramic matrix composite (CMC); Al$_2$O$_3$

---

1. Introduction

The efficiency of a gas turbine is determined by its pressure ratio and its maximum service temperature. A way to increase the efficiency is to increase the maximum temperature. However, the maximum service temperature is limited by the physical and chemical stability of the components in the high-temperature section. A significant temperature increase was achieved by the use of single-crystal super alloys, thermal barrier coatings, and complex cooling systems. Unfortunately, the complex cooling of the components leads to high efficiency losses. Therefore, there is a need for new high-temperature materials with higher temperature capabilities, so that the cooling of the components can be reduced or even omitted [1,2].

Ceramics, such as SiC and Al$_2$O$_3$, offer high temperature and chemical stability and therefore are promising materials for high-temperature applications. Reinforcing of the ceramic matrix with ceramic fibers avoids catastrophic failure due to crack deflection and bridging mechanisms and leads to a pseudo-plastic failure behavior of the ceramic material [3]. These so-called ceramic matrix composites (CMCs) combine high chemical and thermal stability with moderate creep rates and high strength. Among the CMCs, those based on SiC and Al$_2$O$_3$ are most common. Compared to SiC, Al$_2$O$_3$ offers higher resistance against oxidation and corrosion [3]. For this reason, an alumina-based CMC was chosen in this study.

Nevertheless, at temperatures above 1200 °C, the water vapor in the combustion atmosphere causes corrosion reactions and the formation of volatile hydroxides. Environmental barrier coatings (EBCs) are needed to protect the material from degradation [4].

$$Al_2O_3 \text{ (s)} + 3H_2O \text{ (g)} \rightleftharpoons 2Al(OH)_3 \text{ (g)} \tag{1}$$

An EBC should have a high resistance against water vapor at elevated temperatures; therefore, the corrosion rates of many materials were studied in the past (see Fritsch et al. [5–8], Herrmann et al. [9]). Besides the corrosion resistance, the coefficient of thermal expansion (CTE) is a crucial parameter for the performance of an EBC. Large CTE differences between substrate and EBC cause stresses during heating and cooling and may lead to premature failure. Therefore, different materials were categorized according to their corrosion rates and their CTE and presented in Figure 1. For better orientation, data for SiC/SiC and Al_2O_3/Al_2O_3 CMCs were added. It can be seen from Figure 1 that different CMCs require different type of coatings. Rare-earth disilicides are promising coating candidates for SiC/SiC composites [10,11], while rare-earth monosilicides and yttrium aluminates are potential EBC candidates for Al_2O_3/Al_2O_3 CMCs.

Some of these materials have been studied as potential EBC for oxide/oxide CMCs. Among these materials, yttria-stabilized zirconia (YSZ) coatings have been intensively investigated, since YSZ is the material of choice for thermal barrier coatings (TBCs). The suitability of plasma-sprayed, sputtered, and electron-beam physical vapor deposition (EB-PVD) YSZ coatings on wound highly porous oxide (WHIPOX) [12] has been studied by Braue et al. and Mechnich et al. [13–15]. To increase the bonding between substrate and coating, a reaction-bonded alumina (RBAO) [16] bond coat had to be used. Because of the CTE difference between CMC and YSZ, the coating adhesion was quite weak. Furthermore, it has been reported by Vassen et al. [17] that mechanical treatment of CMCs can cause severe damage to the matrix or even its failure.

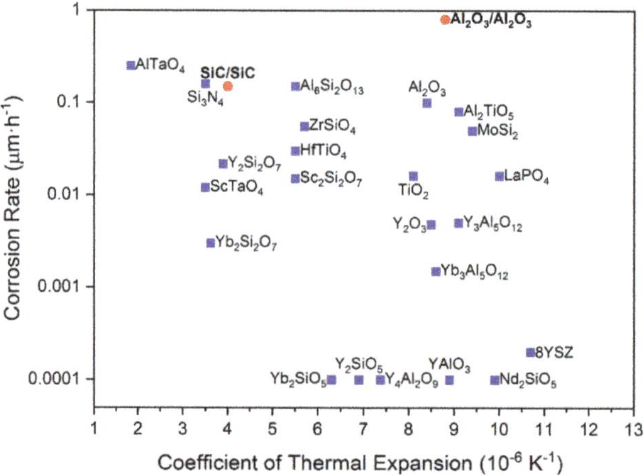

Figure 1. Corrosion rates and coefficient of thermal expansion of selected materials (blue boxes) and ceramic matrix composites (CMCs) (red circles) (data from references [5,7,18]).

Mechnich et al. [19] studied the adhesion of atmospheric plasma-sprayed (APS) Y_2O_3 coatings on WHIPOX substrates that were previously coated with an RBAO bond coat. The coating adhesion was tested by furnace cycling tests at 1200 °C. The results showed a good adhesion and no failure after 500 cycles. SEM investigations revealed the formation of yttrium aluminates at the coating–substrate interface. The formation of this reaction layer is believed to result in a good coating performance.

Atmospheric plasma-sprayed $Gd_2Zr_2O_7$ coatings on an Al_2O_3-based CMC has been studied by Gatzen et al. [20]. Pull adhesion tests showed weak adhesion of $Gd_2Zr_2O_7$ coatings on substrates without pretreatment, while the coating adhesion on laser-structured substrates was significantly increased.

Gerendas et al. [21] published a study on the performance of several APS-coated systems (YSZ, spinel, YSZ/silicate, mullite/silicate) on different oxide/oxide CMCs (UMOX^{TM}, WHIPOX^{TM}, OXIPOL^{TM} (Oxidic CMC based on Polymers)). All substrates were sandblasted before coating application, and in the case of WHIPOX, an RBAO bond coat was additionally applied. The coating performance was tested with burner rig tests, and the results suggested a strong dependence of adhesion on the used CMC material.

All of the above-mentioned coating strategies involve surface preparation by sandblasting, laser structuring, or even the application of a bond coat. Furthermore, the used coatings offer only moderate corrosion rates, but as Figure 1 suggests, there are materials with corrosion rates one or two orders of magnitude lower and with more suitable CTEs. An example is $YAlO_3$ (YAP, yttrium aluminum perovskite), which has a very low corrosion rate and a CTE close to that of the used CMC material. Because of the great chemical similarity of $YAlO_3$ and Y_2O_3 and according to the phase diagram [22], a reaction layer may be formed at the interface ($Y_3Al_5O_{12}$, YAG, yttrium aluminum garnet). Studies on Y_2O_3 coatings [19] have shown that this reaction layer is very beneficial for coating adhesion. Therefore, coatings with good adhesion and long service life are expected. However, there are no studies yet on the suitability of $YAlO_3$ either as TBC or as EBC.

In the past, the production of plasma-sprayed yttrium aluminate coatings (YAG, $Y_3Al_5O_{12}$) proved to be particularly challenging [23]. The challenging behavior of $Y_3Al_5O_{12}$ may also be relevant for $YAlO_3$ coatings. Nevertheless, $YAlO_3$ is a promising EBC candidate. Therefore, plasma-sprayed $YAlO_3$ coatings as EBC for an Al_2O_3-based CMC will be discussed in this study, paying special attention to the phase evolution during plasma spraying.

Because of the expected chemical bonding between coating and substrate, which is believed to cause good coating adhesion and to minimize the processing effort, no additional bond coat or surface treatment was used in this study. The coating adhesion of the plasma-sprayed coatings was studied by means of pull–adhesion tests (PAT) and thermal cycling tests. Furthermore, the resistance of the $YAlO_3$ coatings against CMAS corrosion was tested.

2. Materials and Methods

The CMC used for this study was the commercially available material FW12 (Pritzkow Spezialkeramik, Filderstadt, Germany), consisting of alumina fibers (Nextel 610) that are embedded in a porous matrix of 85% alumina and 15% yttria-stabilized zirconia (YSZ). Before coating application, the samples were ultrasonically cleaned.

The coatings were applied by plasma spraying. The coatings were produced by means of APS and very low pressure plasma spraying (VLPPS). The coating parameters are described in Table 1. For the production of the APS coatings, an Oerlicon Metco multicoat-facility (Wohlen, Switzerland) was used, equipped with a TriplexPro-210 gun that was mounted on a six-axis robot. The VLLPS coating runs were carried out in a vacuum chamber (2 mbar), using an O3CP gun (Oerlicon Metco, Wohlen, Switzerland). The substrate temperatures during processing were measured by IR pyrometry.

Table 1. Used coating parameters for the production of $YAlO_3$ coatings.

Process	Distance (mm)	$T_{Preheat}$ (°C)	$T_{Coating}$ (°C)	Current (A)	Ar (SLPM)	He (SLPM)	Carrier Gas (SLPM)	Passes	Thickness (μm)
APS	70 200	600	750 550	520	50	4	3	2	120
VLPPS	700	1025	1060	2150	110	20	11	2 min	120

SLPM = Standard Liter Per Minute.

In both processes, a spray-dried YAlO$_3$ powder with a particle size of 37 μm and a purity of 96% was used. X-ray diffraction (XRD) measurements (Figure 2) revealed the presence of Y$_4$Al$_2$O$_9$ and Y$_3$Al$_5$O$_{12}$ as secondary phases.

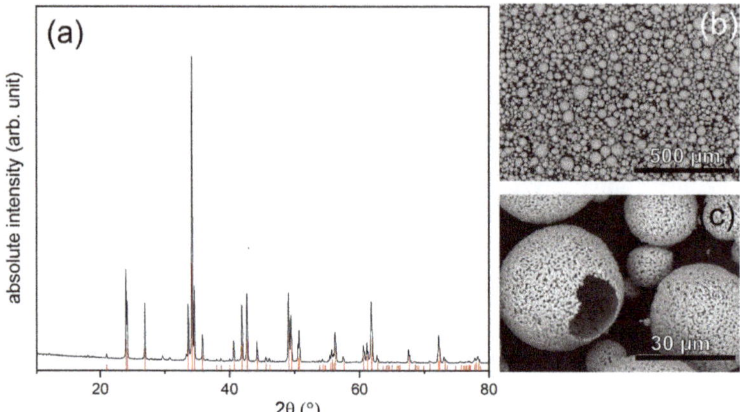

Figure 2. XRD measurement (**a**) and SEM images (**b**,**c**) of the used YAlO$_3$ powder. The diffraction pattern of YAlO$_3$ is shown in red.

After coating manufacturing, the samples were furnace-cycled in air for 4 × 20 h at 1200 °C. In order to assess the thermal and chemical stability of the coatings, XRD measurements were carried out before and after thermal aging. The XRD measurements were carried out with a Bruker D4 Endeavor (Karlsruhe, Germany), using Cu-Kα-radiation (λ = 1.54187 Å). Rietveld refinements were carried out using FullProf suite [24].

The coating adhesion of the as-sprayed coatings was measured with an Elcometer 510 (Aalen, Germany), according to ASTM D4541 [25]. Test dollies with a 10 mm diameter were used and glued to the sample with Araldite two-part epoxy adhesive. Furthermore, the coating performance during thermal cycling was tested by burner rig tests, using the setup described by Traeger et al. [26]. The samples were mounted in a ceramic sample holder and subjected to cycles of 5 min of heating followed by 2 min of cooling. The temperature was measured with pyrometers from the front and back sides. The used temperature program consisted of 510 cycles at 1200 °C, followed by 500 cycles at 1300 °C. The CMAS stability of the samples was tested with a special burner rig, described by Steinke et al. [27]. The surface temperature was set to 1250 °C to ensure melting of the CMAS components.

Materialographic cross sections of the samples before and after testing were prepared. For this, the samples were embedded into resin, cut, and wet-ground with successively finer abrasive paper down to a grit designation of P4000. Afterwards, the samples were polished with diamond suspensions. The polished samples were sputtered with platinum (Leica EM ACE200, Vienna, Austria) and analyzed by scanning electron microscopy (SEM) (Hitachi, TM3000, Tokyo, Japan).

3. Results and Discussion

3.1. Coating Formation

3.1.1. Atmospheric Plasma Spraying

The YAlO$_3$ coatings were produced using APS at varying stand-off distances and resulting different coating temperatures. The XRD-measurements of the APS YAlO$_3$ coatings before and after thermal cycling are presented in Figure 3. The coating sprayed at higher stand-off distance and thus lower substrate temperature was amorphous in the as-sprayed state. The thermal cycling led to the crystallization of the coating. Phase analysis revealed the presence of YAlO$_3$, Y$_3$Al$_5$O$_{12}$, and Y$_4$Al$_2$O$_9$. The refined phase content of the different phases is given in Table 2. Phase analysis showed that the three yttrium aluminates occurred in almost equal fractions.

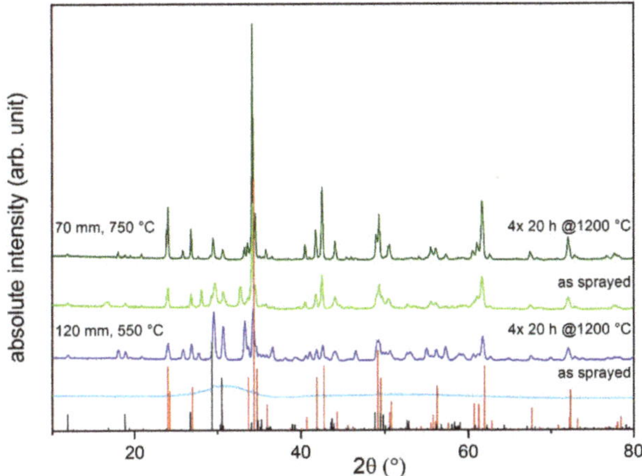

Figure 3. XRD measurements before (light) and after (dark) thermal treatment of APS YAlO$_3$ coatings; theoretical diffraction patterns of Y$_4$Al$_2$O$_9$ (black) and YAlO$_3$ (red).

Table 2. Results of Rietveld refinements of APS YAlO$_3$ coatings.

Phase	70 mm As-Sprayed	70 mm Sintered	120 mm As-Sprayed	120 mm Sintered
YAlO$_3$	74%	90%	–	39%
Y$_4$Al$_2$O$_9$	14%	6%	–	32%
Y$_3$Al$_5$O$_{12}$	12%	4%	–	23%

The coating sprayed at lower stand-off distance and thereby higher substrate temperature was crystalline in the as-sprayed state. The main phase was the desired YAlO$_3$ phase, but significant amounts of Y$_3$Al$_5$O$_{12}$ and Y$_4$Al$_2$O$_9$ were present. The heat treatment led to further crystallization and phase segregation of the coating. The composition shifted in favor of the desired YAlO$_3$ phase.

To enable crystallization of the YAlO$_3$ coating, a certain energy level must be exceeded. Furthermore, the cooling rate should be slow, so that there is enough time for atomic rearrangement and the formation of a long-range order. The cooling rate can be lowered by using higher plasma power, by shortening the stand-off distance, or by substrate preheating. The splat solidification time [28] is a key factor. A dependence of the splat solidification time on the temperature gradient between splat and substrate has been reported [28,29]. A decreased gradient leads to a longer solidification time, consequently the splats deposited at higher temperatures have more time for crystallization. This explains why the YAlO$_3$ coatings, sprayed at shorter distances (70 mm) and higher substrate temperatures (750 °C) were crystalline, while those sprayed at 120 mm distance and a significantly lower temperature (550 °C) were amorphous.

The formation of secondary phases can be explained by the evaporation of elements during thermal spraying, causing a shift in the phase diagram [23]. In this case, evaporation of alumina might occur, since alumina has a higher vapor pressure then yttria [30]. A more probable explanation is demixing due to rapid quenching of splats and impure base material (see Figure 2). This is supported by the fact that the coating, which was sprayed at higher distance and lower substrate temperature resulting in a larger temperature gradient between splat and substrate, differed significantly more from the desired phase composition. Furthermore, yttrium aluminates with both high alumina content and low-alumina content were found. If evaporation of alumina was the main mechanism, the phase equilibrium would be shifted to the phase with low alumina content ($Y_4Al_2O_9$).

SEM images of cross sections of $YAlO_3$ coatings before and after thermal cycling are presented in Figure 4. The amorphous coatings sprayed at higher distances (120 mm) and thus lower substrate temperatures were relatively dense and showed only few cracks. The presence of secondary phases, which was already revealed by XRD-measurements, can also be observed in the SEM images. The SEM images reveal that the coating was delaminated during thermal cycling. Coating delamination might be attributed to the occurring crystallization of the coating.

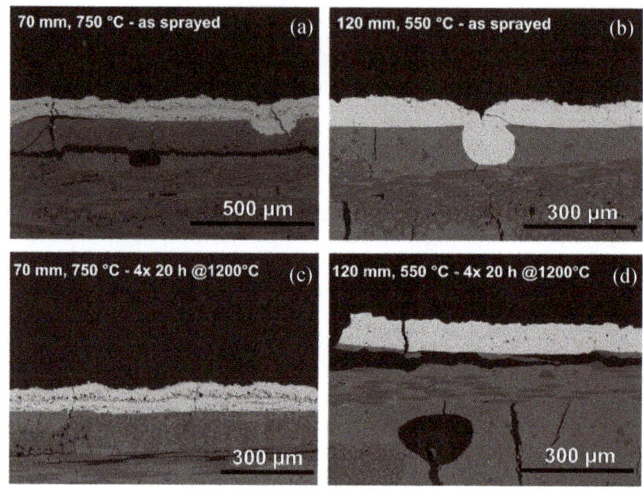

Figure 4. SEM images of the APS $YAlO_3$ coatings before (Coatings sprayed at a distance of 70 mm (**a**) and 120 mm (**b**)) and after thermal treatment (Coatings sprayed at a distance of 70 mm (**c**) and 120 mm (**d**)).

The coatings obtained at a smaller stand-off distance (70 mm) were already crystalline in the as-sprayed state. However, large pores were present in these coatings, especially at the interface between the different coating layers. Sporadic delamination was observed even in the as-sprayed state, and the coatings were prone to form segmentation cracks. Both cracks and pores have a negative effect on corrosion resistance, as they increase the permeability of the EBC to water vapor. Here, the segmentation cracks seemed to continue into the substrate, which could additionally affect the structural integrity of the material.

These results are in good agreement with the results for APS $Y_3Al_5O_{12}$ coatings of Weyant et al. [24]. A systematic investigation of the effects of several spraying parameters on the resulting coating microstructure led to the conclusion that the used power and the stand-off distance are the determining factors. Unfortunately, increasing spraying distances led to both lower porosity and reduced crystallinity. Moreover, increased power caused increased crystallinity, but the porosity was increased as well [24]. The same behavior was found in this study for atmospheric plasma-sprayed $YAlO_3$ coating:

The formation of pores seemed to go along with the crystallization of the as-sprayed $YAlO_3$ phase. This can be attributed to the differences in density between melt and solid. It is assumed that the density of amorphous $YAlO_3$ is close to that of the liquid phase, which is slightly below 4.0 g·cm^{-3} [31–33]. The density of the crystalline phase is reported to be significantly higher, at 5.35 g·cm^{-3} [34]. These differences lead to shrinkage during crystallization and therefore cause the formation of large pores and cracks. In the case of initially dense but amorphous coatings, the tensile stresses that occur during the crystallization processes resulting from the thermal treatment are high, and as a consequence, the coating is delaminated.

The formation of a reaction layer, as reported for Y_2O_3 coatings, was not observed. It is assumed that the stresses occurring in these coatings were so high that delamination occurred before the reaction could take place. Furthermore, poor wetting could also be a reason for the absence of a reaction layer.

3.1.2. Very Low Pressure Plasma Spraying

Coatings were produced with VLPPS, in order to overcome the aforementioned problems related to crystallization in the APS $YAlO_3$ coatings. VLPPS offers high power, high temperatures and low cooling rates. The results of the VLPPS $YAlO_3$ coatings are presented in Figure 5.

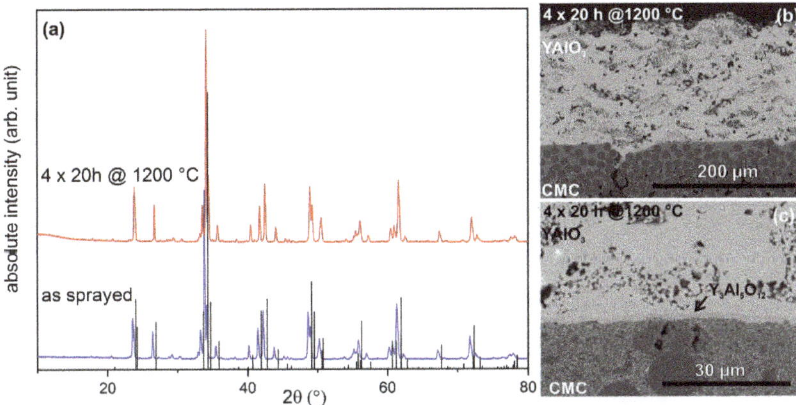

Figure 5. (**a**): XRD-measurement of the VLPPS $YAlO_3$ coatings in the as-sprayed and thermally treated state. (**b**,**c**): SEM images of the VLPPS $YAlO_3$ coatings after thermal treatment.

The XRD measurements showed that the $YAlO_3$ coatings were crystalline directly after spraying. The major phase was $YAlO_3$, and only very weak reflections of a secondary phase (yttrium aluminum monoclinic, YAM, $Y_4Al_2O_9$) were visible. Due to a slight bending of the coated sample, the measured XRD signals were slightly offset from the theoretical ones. The XRD measurement of the thermally treated sample showed hardly any deviations from the original XRD. This shows that there were no phase transformations or significant crystallization processes during heating/cooling, which could cause stresses and cracks in the coating. The formation of nearly single-phase $YAlO_3$ can be attributed to the lower cooling rate. The increased surface temperature led to an increased solidification time. Furthermore, the low chamber pressure led to significantly slower cooling rates of the coating and substrate. As a consequence, the coating had more time to crystallize. Therefore, the resulting coating had a high crystallinity and a high content of the desired $YAlO_3$ phase. Rietveld refinements led to a phase content >88% of the desired $YAlO_3$ phase. This represents a significant increase compared to the coatings produced under atmospheric conditions.

The SEM images of the cross sections of the VLPPS $YAlO_3$ coatings are shown in Figure 5a. The coating had a finely distributed porosity, which was caused by not fully molten particles. The pores were distributed homogenously in the coating. In contrast to APS coatings, no vertical cracks were observed. The VLPPS $YAlO_3$ coatings formed a reaction layer at the interface with the substrate. This reaction zone can be observed even in the as-sprayed samples (Figure 6). This implies that the energy needed for the reaction [35] was reached in these experiments. The higher temperatures might also increase the wetting of the splats and therefore increase the contact between coating and substrate, leading to the formation of a 1–2 µm-thick reaction zone. Because of the position of $YAlO_3$ in the Y_2O_3–Al_2O_3 phase diagram [36], the formation of $Y_3Al_5O_{12}$ is likely. In contrast to the results for the APS coatings, there was no delamination within the coating–substrate system after heat treatment.

Due to the high crystallinity and the formation of chemical bonding between coating and substrate, a strong coating adhesion is expected.

Figure 6. SEM image of the VLPPS YAlO$_3$ coating in the as-sprayed state, with a Y$_3$Al$_5$O$_{12}$ reaction layer at the coating substrate interface.

3.2. Pull–Adhesion Tests

The coating adhesion was tested with pull–adhesion tests, and the results are illustrated in Figure 7. The adhesion of the APS coatings was very poor, and both coatings failed directly at the beginning of the test. High stresses in the coatings might be the reason for the bad coating adhesion. The adhesion of the VLPPS coatings was significantly higher. The measured strength was in the order of that of the CMC itself.

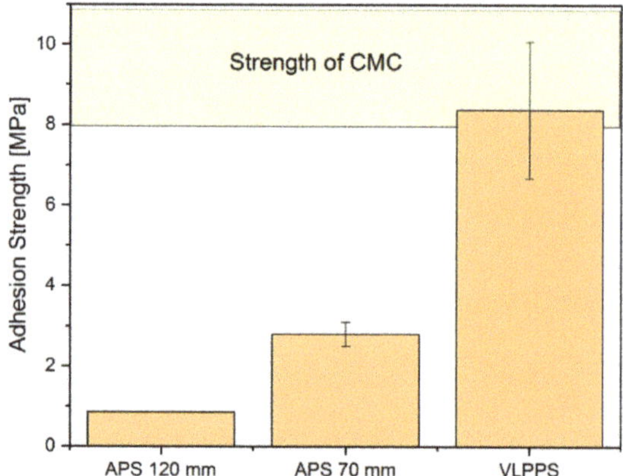

Figure 7. Measured adhesion strength of as-sprayed APS and VLPPS YAlO$_3$ coatings.

The samples were embedded and cut after adhesion testing. SEM images of the polished cross sections are shown in Figure 8. The high stresses in the APS coatings led to failure partly within the CMC and partly at the coating–substrate interface. It is assumed that slight reactions between coating and substrate or clamping occurred, which caused a partial failure within the matrix of the CMC.

The good adhesion between the VLPPS coating and the CMC caused a failure deep within the substrate itself. This is consistent with the measured adhesion strength, as the measured adhesion strength was in the order of the strength of the CMC itself. The high adhesion strength was attributed to the occurrence of a chemical reaction at the coating–substrate interface.

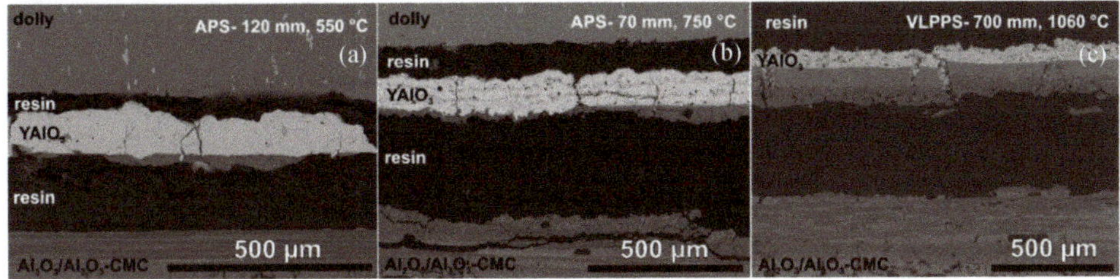

Figure 8. SEM images of the APS YAlO$_3$ coatings sprayed at 120 mm (**a**) and 70 mm (**b**) and of the VLPPS YAlO$_3$ coating after pull–adhesion tests PAT (**c**).

3.3. Burner Rig Tests

The pull–adhesion tests of the VLPPS YAlO$_3$ coatings revealed a promising adhesion strength. On the basis of previous investigations on Y$_2$O$_3$ coatings resulting in excellent stability due to the formation of a reaction layer at the interface [20], a high thermal cycling lifetime of the VLPPS coatings was expected. The APS coatings showed poor adhesion strength, and as a consequence, these coatings were excluded from the following investigations.

The thermal cycling tests were stopped after 1010 cycles (510 cycles at 1200 °C and 500 cycles at 1300 °C), which corresponds to the upper limit of the average lifetime of standard TBC samples in this test rig. During cycling, no macroscopic failure occurred (see Figure 9). SEM images of the VLPPS YAlO$_3$ coatings after 1010 cycles are shown in Figure 10. The coating still shows a fine distributed porosity. The Y$_3$Al$_5$O$_{12}$ containing reaction zone at the coating substrate interface was about 1–2 µm-thick. No growth took place within the duration of the test.

Figure 9. Photograph of VLPPS YAlO$_3$ coating on FW12 after 1010 cycles of burner rig testing.

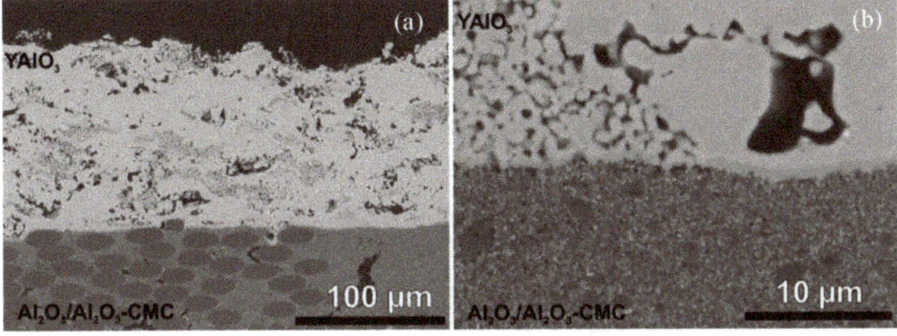

Figure 10. SEM images of cross sections of VLPPS YAlO$_3$ coatings after 1010 cycles of burner rig testing (**a**), close up of the coating substrate interface (**b**).

No delamination or crack formation was observed after 1010 cycles of burner rig testing; therefore, a significantly longer lifetime can be expected. The long cycling lifetime of the YAlO$_3$ coatings on the FW12-CMC is attributed to the strong bond between the coating and the substrate due to the formation of a thermodynamically stable reaction layer at the interface.

3.4. CMAS Tests

The coating's resistance to CMAS corrosion was also investigated. The tests were stopped after 274 cycles, which corresponds to about twice the average lifetime of a typical YSZ coating in this rig. After 274 cycles, no macroscopic coating failure was observed (see Figure 11). The XRD measurements of the sample before and after the CMAS test are shown in Figure 12. Besides signals from the CMAS and the coating itself, the formation of calcium–yttrium silicate oxyapatite (Ca$_2$Y$_8$(SiO$_4$)$_6$O$_2$) as a reaction product between YAlO$_3$ and CMAS was observed. This phase was already observed for other Y-containing coatings [37,38]. In addition, the formation of Y-depleted yttrium aluminate Y$_3$Al$_5$O$_{12}$ was observed accordingly. Turcer et al. [38] recently reported the excellent CMAS resistance of dense YAlO$_3$ pellets due to the fast formation of a protective Ca–Y–Si apatite phase. Our results are in good agreement with those from Turcer et al.

Figure 11. Photograph of VLPPS YAlO$_3$ coating on FW 12 after 274 cycles in CMAS test rig.

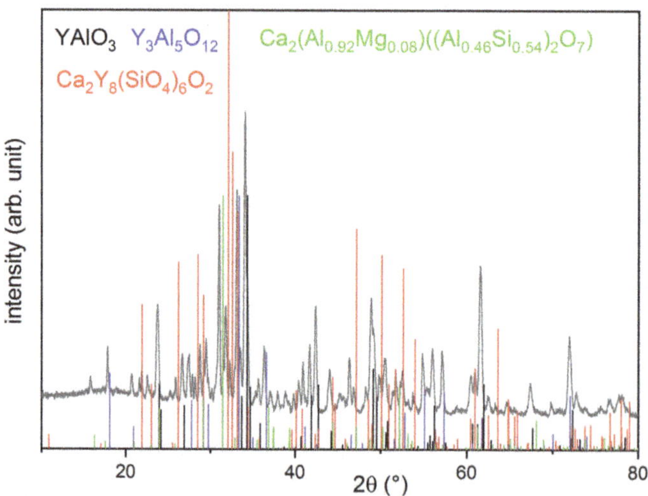

Figure 12. Measured XRD (grey) of VLPPS YAlO$_3$ coating after 274 cycles in CMAS test and theoretical diffraction patterns of YAlO$_3$ (black), Y$_3$Al$_5$O$_{12}$ (blue), Ca$_2$(Al$_{0.92}$Mg$_{0.08}$)((Al$_{0.46}$Si$_{0.54}$)$_2$O$_7$) (green), and Ca$_2$Y$_8$(SiO$_4$)$_6$O$_2$ (red).

The SEM images of the resulting cross sections are presented in Figure 13. No indications for crack formation or beginning delamination were found. Because of the porosity of the sample, the molten CMAS was able to infiltrate the upper part of the VLPPS YAlO$_3$ coating (about 46 µm). The SEM images and the corresponding EDS measurements show the sharp borderline of the infiltrated zone. The reaction zone between CMAS and YAlO$_3$ can be clearly seen. The CMAS infiltration and the following reactions led to densification of the infiltrated parts of the coating. The densification of the upper coating may hinder further infiltration and thus extend the lifetime of the coating. Furthermore, the CTE of the CMAS constituents is in the area of 9.7×10^{-6} K^{-1} [39,40] and close to the CTE of YAlO$_3$ (8.9×10^{-6} K^{-1} [41]); thus, low stresses were expected during testing. Because of the remaining porosity of the lower coating parts, occurring stresses can be relaxed. As a consequence, even after 274 cycles, no spallation occurred.

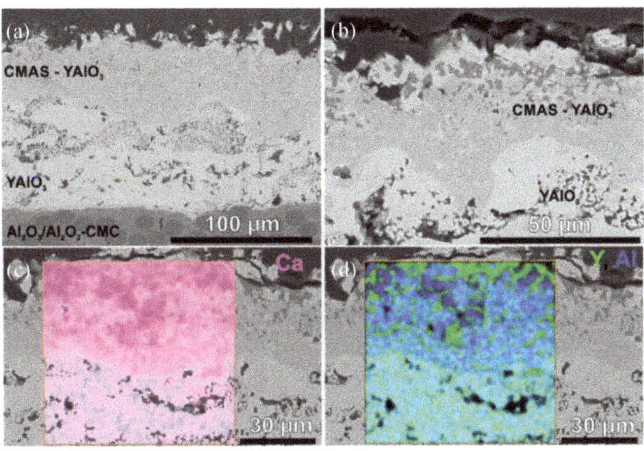

Figure 13. SEM images of the VLPPS YAlO$_3$ coatings after 274 cycles in the CMAS test rig (**a,b**). EDS mappings for Ca (**c**) and Y and Al (**d**).

4. Conclusions

The suitability of YAlO$_3$ coatings as EBC for an Al$_2$O$_3$/Al$_2$O$_3$–CMC was examined. YAlO$_3$ coatings were produced by means of APS and VLPPS, without applying any bond coats. Furthermore, no surface pretreatment was carried out. The APS YAlO$_3$ coating trials resulted in coatings with either poor crystallinity or large pores, both of which are not favorable for an EBC.

YAlO$_3$ coatings produced by VLPPS offered high crystallinity and purity. The VLPPS coatings showed strong adhesion, which was attributed to the formation of chemical bonding between coating and substrate. This thermodynamically stable reaction layer at the coating–substrate interface, probably consisting of Y$_3$Al$_5$O$_{12}$, was found directly after spraying.

The VLPPS YAlO$_3$ were chosen for further investigations. The thermal cycling lifetime was tested with burner rig tests. The coatings passed the test consisting of 1010 cycles without failure. Furthermore, the hot corrosion behavior was tested with CMAS tests. The coatings showed excellent CMAS resistance, by withstanding more than 274 cycles without failure. The high cycling lifetime was attributed to the densification of the coating due to CMAS infiltration and formation of a reaction product, blocking further infiltration.

The use of VLPPS for the production of YAlO$_3$ coatings led to stable coatings with a microstructure with intermixed porous and dense zones. This led to a high thermal cycling lifetime and high CMAS resistance. Furthermore, the formation of a reaction zone at the interface enabled good coating adhesion. Therefore, VLPPS YAlO$_3$ coatings are promising candidates for EBCs for Al$_2$O$_3$-based CMCs, as they offer high adhesion strength, high thermal stability, and high corrosion resistance.

Author Contributions: Conceptualization and methodology, C.G. and D.E.M.; investigation, C.G.; writing—original draft preparation, C.G.; writing—review and editing, D.E.M., R.V. and O.G.; supervision, D.E.M. and R.V.; project administration, R.V.

Funding: This research received no external funding.

Acknowledgments: The authors would like to thank Ralf Laufs, Frank Kurze, and Karl-Heinz Rauwald for their assistance during plasma spraying. The authors acknowledge Doris Sebold for the SEM analysis. In addition, tha authors would like to thank Volker Bader and Martin Tandler for the execution of the heat treatments, burner rig tests, and CMAS tests.

Conflicts of Interest: The authors declare no conflict of interest.

References

1. Perepezko, J.H. The hotter the engine, the better. *Science* **2009**, *326*, 1068–1069. [CrossRef] [PubMed]
2. Xin, Q. *Diesel Engine System Design*; Woodhead Publishing: Oxford, UK, 2013; pp. 860–908.
3. Krenkel, W. *Keramische Verbundwerkstoffe*; WILEY-VCH Verlag GmbH & Co. KGaA: Weinheim, Germany, 2003.
4. Opila, E.J.; Myers, D.L. Alumina volatility in water vapor at elevated temperatures. *J. Am. Ceram. Soc.* **2004**, *87*, 1701–1705. [CrossRef]
5. Fritsch, M. *Heißgaskorrosion Keramischer Werkstoffe in H2O-Haltigen Rauchgasatmosphären*; Fraunhofer IRB Verlag, TU Dresden: Dresden, Germany, 2007.
6. Fritsch, M.; Klemm, H. The water vapor hot gas corrosion of MGC materials with Al_2O_3 as a phase constituent in a combustion atmosphere. *J. Eur. Ceram. Soc.* **2008**, *28*, 2353–2358. [CrossRef]
7. Fritsch, M.; Klemm, H. The water-Vapour hot gas corrosion behavior of Al_2O_3-Y_2O_3 materials, Y_2SiO_5 and $Y_3Al_5O_{12}$-Coated alumina in A combustion environment. *Adv. Ceram. Eng. Sci. Proc.* **2006**, *27*, 149–159.
8. Fritsch, M.; Klemm, H.; Herrmann, M.; Schenk, B. Corrosion of selected ceramic materials in hot gas environment. *J. Eur. Ceram. Soc.* **2006**, *26*, 3557–3565. [CrossRef]
9. Herrmann, M.; Klemm, H. Corrosion of ceramic materials. *Compr. Hard Mater.* **2014**, *2*, 413–446.
10. Bakan, E.; Marcano, D.; Zhou, D.; Sohn, Y.J.; Mauer, G.; Vaßen, R. $Yb_2Si_2O_7$ environmental barrier coatings deposited by various thermal spray techniques: A preliminary comparative study. *J. Therm. Spray Technol.* **2017**, *26*, 1011–1024. [CrossRef]
11. Richards, B.T.; Wadley, H.N.G. Plasma spray deposition of tri-layer environmental barrier coatings. *J. Eur. Ceram. Soc.* **2014**, *34*, 3069–3083. [CrossRef]
12. Göring, J.; Kanka, B.; Schmücker, M.; Schneider, H. A Potential Oxide/Oxide Ceramic Matrix Composite for Gas Turbine Applications. In Proceedings of the ASME Turbo Expo 2003, collocated with the 2003 International Joint Power Generation Conference, Atlanta, GA, USA, 16–19 June 2003; pp. 621–624.
13. Braue, W.; Mechnich, P. Tailoring protective coatings for all-oxide ceramic matrix composites in high temperature-/high heat flux environments and corrosive media. *Mater. Werkst.* **2007**, *38*, 690–697. [CrossRef]
14. Braue, W.; Mechnich, P. *Schutzschichtkonzepte für Oxidische Faserverbundwerkstoffe*; DLR Kolloquium: Köln, Germany, 2004.
15. Singh, D.; Zhu, D.; Zhou, Y.; Singh, M. *Design, Development, and Applications of Engineering Ceramics and Composites: Ceramic Transactions*; John Wiley & Sons: Hoboken, NJ, USA, 2010.
16. Mechnich, P.; Braue, W.; Schneider, H. Multifunctional reaction-bonded alumina coatings for porous continuous fiber-reinforced oxide composites. *Int. J. Appl. Ceram. Technol.* **2004**, *1*, 343–350. [CrossRef]
17. Vaßen, R.; Bakan, E.; Gatzen, C.; Kim, S.; Mack, D.E.; Guillon, O. Environmental barrier coatings made by different thermal spray technologies. *Coatings* **2019**. in Preparation.
18. Fritsch, M.; Klemm, H.; Herrmann, M.; Michaelis, A.; Schenk, B. The water vapour hot gas corrosion of ceramic materials. *Ceram. Forum Int.* **2010**, *87*, 11–12.
19. Mechnich, P.; Braue, W. Air plasma-Sprayed Y_2O_3 coatings for Al_2O_3/Al_2O_3 ceramic matrix composites. *J. Eur. Ceram. Soc.* **2013**, *33*, 2645–2653. [CrossRef]
20. Gatzen, C.; Mack, D.E.; Guillon, O.; Vaßen, R. Surface roughening of Al_2O_3/Al_2O_3-ceramic matrix composites by nanosecond laser ablation prior to thermal spraying. *J. Laser Appl.* **2019**, *31*, 022018. [CrossRef]
21. Gerendás, M.; Cadoret, Y.; Wilhelmi, C.; Machry, T.; Knoche, R.; Behrendt, T.; Aumeier, T.; Denis, S.; Göring, J.; Koch, D.; et al. Improvement of Oxide/Oxide CMC and Development of combustor and turbine components in the HiPOC program. *ASME Turbo Expo* **2011**, *1*, 477–490.

22. Fabrichnaya, O.; Seifert, H.J.; Weiland, R.; Ludwig, T.; Aldinger, F.; Navrotsky, A. Phase equilibria and thermodynamics in the Y_2O_3–Al_2O_3–SiO_2 system. *Z. Metallkd.* **2001**, *92*, 1083–1097.

23. Weyant, C.M.; Faber, K.T. Processing–microstructure relationships for plasma-sprayed yttrium aluminum garnet. *Surf. Coat. Technol.* **2008**, *202*, 6081–6089. [CrossRef]

24. Rodríguez-Carvajal, J. Recent advances in magnetic structure determination by neutron powder diffraction. *Physica B* **1993**, *192*, 55–69. [CrossRef]

25. *ASTM D4541-17, Standard Test Method for Pull-Off Strength of Coatings Using Portable Adhesion Testers*; ASTM International: West Conshohocken, PA, USA, 2017. [CrossRef]

26. Traeger, F.; Vaßen, R.; Rauwald, K.H.; Stöver, D. Thermal cycling setup for testing thermal barrier coatings. *Adv. Eng. Mater.* **2003**, *5*, 429–433. [CrossRef]

27. Steinke, T.; Sebold, D.; Mack, D.E.; Vaßen, R.; Stöver, D. A novel test approach for plasma-Sprayed coatings tested simultaneously under CMAS and thermal gradient cycling conditions. *Surf. Coat. Technol.* **2010**, *205*, 2287–2295. [CrossRef]

28. Heimann, R.B. *Plasma-Spray Coating*; Weinheim VCH: New York, NY, USA, 1996.

29. Chandra, S.; Fauchais, P. Formation of solid splats during thermal spray deposition. *J. Therm. Spray Technol.* **2009**, *18*, 148–180. [CrossRef]

30. Schulz, U.; Saruhan, B.; Fritscher, K.; Leyens, C. Review on advanced EB-PVD ceramic topcoats for TBC applications. *Int. J. Appl. Ceram. Technol.* **2004**, *1*, 302–315. [CrossRef]

31. Aasland, S., McMillan, P.F. Density-driven liquid–liquid phase separation in the system Al_2O_3–Y_2O_3. *Nature* **1994**, *369*, 633–636. [CrossRef]

32. Srivastava, A.K. *Oxide Nanostructures: Growth, Microstructures, and Properties*; Pan Stanford Publishing: Singapore, 2014.

33. Watanabe, Y.; Masuno, A.; Inoue, H. Glass formation of rare earth aluminates by containerless processing. *J. Non-Cryst. Solids* **2012**, *358*, 3563–3566. [CrossRef]

34. Ching, W.Y.; Xu, Y.N. Nonscalability and nontransferability in the electronic properties of the Y–Al–O system. *Phys. Rev. B* **1999**, *59*, 12815–12821. [CrossRef]

35. Fabrichnaya, O.; Seifert, H.J.; Ludwig, T.; Aldinger, F.; Navrotsky, A. The assessment of thermodynamic parameters in the Al_2O_3–Y_2O_3 system and phase relations in the Y–Al–O system. *Scand. J. Metall.* **2001**, *30*, 175–183. [CrossRef]

36. Medraj, M.; Hammond, R.; Parvez, M.A.; Drew, R.A.L.; Thompson, W.T. High temperature neutron diffraction study of the Al_2O_3–Y_2O_3 system. *J. Eur. Ceram. Soc.* **2006**, *26*, 3515–3524. [CrossRef]

37. Ahlborg, N.L.; Zhu, D. Calcium–Magnesium aluminosilicate (CMAS) reactions and degradation mechanisms of advanced environmental barrier coatings. *Surf. Coat. Technol.* **2013**, *237*, 79–87. [CrossRef]

38. Turcer, L.R.; Krause, A.R.; Garces, H.F.; Zhang, L.; Padture, N.P. Environmental-Barrier coating ceramics for resistance against attack by molten calcia-magnesia-aluminosilicate (CMAS) glass: Part I, $YAlO_3$ and γ-$Y_2Si_2O_7$. *J. Eur. Ceram. Soc.* **2018**, *38*, 3905–3913. [CrossRef]

39. Bansal, N.P.; Choi, S.R. *Properties of Desert Sand and CMAS Glass*; NASA Glenn Research Center: Cleveland, OH, USA, August 2014.

40. Reddy, A.A.; Goel, A.; Tulyaganov, D.U.; Kapoor, S.; Pradeesh, K.; Pascual, M.J.; Ferreira, J.M.F. Study of calcium–magnesium–aluminum–silicate (CMAS) glass and glass-ceramic sealant for solid oxide fuel cells. *J. Power Sources* **2013**, *231*, 203–212. [CrossRef]

41. Chaix-Pluchery, O.; Chenevier, B.; Robles, J.J. Anisotropy of thermal expansion in $YAlO_3$ and $NdGaO_3$. *Appl. Phys. Lett.* **2005**, *86*, 251911. [CrossRef]

 *coatings*

Article

Delayed Formation of Thermally Grown Oxide in Environmental Barrier Coatings for Non-Oxide Ceramic Matrix Composites

Hagen Klemm *, Katrin Schönfeld and Willy Kunz

Fraunhofer Institute for Ceramic Technologies and Systems IKTS, Dresden 01277, Germany;
katrin.schoenfeld@ikts.fraunhofer.de (K.S.); willy.kunz@ikts.fraunhofer.de (W.K.)
* Correspondence: hagen.klemm@ikts.fraunhofer.de; Tel.: +49-351-2553-7553

Received: 22 November 2019; Accepted: 17 December 2019; Published: 19 December 2019

Abstract: The oxidation and corrosion behavior at elevated temperatures of a $SiC_F/SiC(N)$ composite with two plasma-sprayed environmental barrier coating (EBC) systems were studied. After both processes, the formation of a silica-based thermally grown oxide (TGO) layer was observed. The formation of this TGO caused two principal failure mechanisms of the EBC. Firstly, spallation of the EBC induced by stresses from volume expansion and phase transformation to crystalline SiO_2 was observed. Water vapor corrosion of the TGO with gap formation in the top region of the TGO was found to be a second failure mechanism. After a burner rig test of the Al_2O_3-YAG EBC system, this corrosion process was observed at the TGO surface and in the volume of the Al_2O_3 bond coat. In the case of the second system, $Si-Yb_2Si_2O_7/SiC-Yb_2SiO_5$, the formation of the TGO could be delayed by introducing an additional intermediate layer based on $Yb_2Si_2O_7$ filled with SiC particles. The SiC particles in the intermediate layer were oxidized and served as getter to reduce the permeation of oxidants (O_2, H_2O) into the material. In this way, the formation of the TGO and the failure mechanisms caused by their formation and growth could be delayed.

Keywords: environmental barrier coatings; non-oxide ceramic matrix composites; oxidation; water vapor corrosion; thermally grown oxide; damage mechanisms

---

1. Introduction

Silicon carbide fiber-based ceramic matrix composites (CMCs) offer a high potential for applications as structural components in advanced gas turbines. In comparison to metallic super alloys, used in the state of the art, the main advantages of these materials were found to be their low specific weight in combination with a superior potential at elevated temperatures up to 1400 °C. Furthermore, among ceramic materials, CMCs are characterized by a damage-tolerant fracture behavior, suggesting them as promising candidates for gas turbine applications as well. During recent years, significant progress has been achieved in material development and processing. However, there are still considerable deficits at present, especially in the long-term behavior of the composites in hot gas atmosphere. Corrosion processes were observed, caused by the high water vapor pressure in combination with high temperatures and gas velocities. The resulting microstructural and mechanical degradation of the composites and the damage mechanisms of these processes have been described in several studies [1–6]. Volatilization of the protective silica-based surface layer by the formation and evaporation of silicon hydroxides ($Si(OH)_4$) was found to be the main process leading to considerable material loss with recession rates in the range of 1 μm/h (Equation (1)). Additional material degradation as a consequence of oxidation processes inside the composite was observed.

$$SiO_2 + 2H_2O_{(g)} = Si(OH)_{4(g)}. \tag{1}$$

Environmental barrier coatings (EBCs) have been the solution to prevent the surface corrosion of the ceramic materials in gas turbine atmospheres [7,8]. During the last few years, different EBC systems have been introduced [8–10]. As a consequence of the complex conditions during operation at elevated temperatures in a hot gas atmosphere, multilayer coatings with special functions were proposed. In this way, several features required to guarantee the long-term stability of the EBC system in hot gas conditions could be realized.

Beside their stability against erosion, interaction with Ca-Mg-Al-silicates (CMAS), or foreign object damage, the top layer of the system must primarily exhibit a superior water vapor corrosion stability. During recent years, several oxide systems with superior corrosion stability have been suggested to protect non-oxide ceramics or CMCs based on Si_3N_4 or SiC against water vapor corrosion [11,12]. Among these oxide systems, rare-earth (RE) silicates have been identified as promising EBC candidates for top layer materials. While the RE-monosilicates are mostly stable in a hot gas environment, the disilicates were found to be partially volatilized with the formation of silicon hydroxide and the more stable monosilicate (Equation (2)) [8,12–17].

$$RE_2Si_2O_7 + 2H_2O_{(g)} = RE_2SiO_5 + Si(OH)_{4(g)} \tag{2}$$

This corrosion process led consequently to the formation of a stable monosilicate layer, influencing the corrosion behavior of the EBC system during long-term application.

Currently, a layer based on metallic silicon is used as a very effective bond coat in several EBC systems [18,19]. With a melting point of about 1410 °C, silicon bond coats are limited in their temperature potential. For use at lower temperatures, however, they are characterized by several benefits. First, the coatings agree well with the coefficient of thermal expansion (CTE) of the non-oxide CMC substrate material. The second point is the getter function of the silicon against the permeation of oxidizing compounds (O_2, H_2O) and prevention of oxidation processes inside the CMC component.

Various multilayer EBC systems with Si bond coats and RE-top coats were demonstrated to be quite effective in the protection of SiC_F/SiC CMC [10,16,17,20,21]. However, during operation at elevated temperatures in hot gas atmosphere, several processes led to degradation of the whole system. A summary of possible failure modes was reported by Lee [19]. Processes like the formation of stresses during thermal cycling, foreign objects, phase transformation, or sintering processes resulted in cracking, delamination, and spallation of the EBC system. Additional chemical processes like water vapor or CMAS corrosion limit the stability and functionality of the protecting system. During long-term use, oxidation processes were found to be an additional critical factor for the stability of the EBC. Diffusion of oxygen and, especially, the permeation of H_2O through the different EBC layers are responsible for the formation of a thermally grown oxide (TGO) layer of mainly silica at the upper side of the Si bond coat. With growing thickness of the SiO_2 TGO layer, crystallization and phase transition processes (cristobalite) were observed, finally leading to stresses in the EBC system with the consequence of cracking and spallation of the EBC [10,16,22].

In real conditions, the formation of the TGO layer cannot be avoided. There will always be permeation of oxygen and water into the material, finally leading to oxidation processes inside. However, there are strategies to minimize the rate of TGO layer formation and their following influences. First, the transport of the oxidants (O_2 and H_2O) through the EBC system should be considered. There is still a considerable lack of data about the permeation properties (diffusion coefficient, oxidant solubility) of the various materials used in the EBC layer system. Furthermore, the morphology of the different layers, like layer thickness, porosity, crystallinity, and grain boundary structure, has to be studied regarding their influence on their permeation properties. Recently, an example in this direction was introduced by Lee [19]. The TGO growth rate in an EBC system with the Si bond coat and $Yb_2Si_2O_7$ was found to be significantly reduced by modifying the $Yb_2Si_2O_7$ layer with various oxides (Al_2O_3, mullite, $Y_3Al_5O_{12}$). As a conclusion of these results, he suggested a modification of the SiO_2 network structure of the TGO by incorporating Al^{3+}- and Yb^{3+}-ions, consequently leading to lower permeation rates of oxidants (H_2O) through the TGO layer.

A defined control of the oxidation and corrosion processes itself was found to be a second strategy. This can be performed by modification of the oxidation mechanism, e.g., defined reaction products or the location where the oxidation takes place. An example for such a strategy was reported for monolithic Si_3N_4 with SiC or $MoSi_2$ additions [12,23]. During oxidation of these composites, a changed oxidation mechanism with the formation of Si_2ON_2 in the top region of the bulk material was observed, leading finally to less defect formation caused by the oxidation processes in the microstructure of the material. The focus of this study was placed in a similar direction, namely, to avoid the formation of a TGO as a reaction layer at, e.g., the silicon bond coat, by a defined reaction of the permeating oxidants at other regions.

2. Materials and Methods

The base CMC material was fabricated by winding technology with polycrystalline SiC fibers, Tyranno SA3 (UBE Industries, Tokyo, Japan). Prior to winding, the desized SiC tows were infiltrated with an aqueous slurry composed of SiC powder, Sintec 15 (Saint Gobain, Courbevoie, France), and 20 vol.% sintering additives with Al_2O_3, AKP 50 (Sumitomo Chemical, Tokyo, Japan); Y_2O_3, Grade C (H.C. Stark, Goslar, Germany); and SiO_2, Aerosil Ox 50 (Evonic Industries, Essen, Germany). The wound cylinder (85° winding angle) was cut and pressed, opening into a flat sheet. Matrix formation was performed in five steps of precursor infiltration and pyrolysis (PIP) with commercially available polysilazane Si-C-N precursor, HTT 1800 (Clariant Advanced Materials GmbH, Muttenz, Switzerland). Afterward, the composite was sintered at 1400 °C in nitrogen atmosphere. Finally, a $SiC_F/SiC(N)$ composite with a fiber volume content between 40% and 50% and an open porosity of about 10% was obtained. Further details about the CMC fabrication are described in [24]. Bars with dimensions of $3 \times 10 \times 36$ mm^3 were used as test samples.

The first EBC system was a bond coat from Al_2O_3 with a top coat of yttrium aluminum garnet ($Y_3Al_5O_{12}$, YAG). Both layers were fabricated by atmospheric plasma spraying. The second system was a three-layer coating system with a Si bond coat, an intermediate layer consisting of a mixture of $Yb_2Si_2O_7/SiC$ and Yb_2SiO_5 as the top coat. While the Si bond coat was fabricated by atmospheric plasma spraying (APS), the two rare-earth-containing layers were fabricated by suspension plasma spraying (SPPS). An overview of the coatings fabricated is given in Table 1:

Table 1. Average coating thickness of the layers in the two EBC systems.

	Thickness/μm	Fabrication
Al_2O_3-YAG		
Al_2O_3	50 to 70	APS
YAG	80 to 120	APS
Si-$Yb_2Si_2O_7/SiC$-Yb_2SiO_5		
Si	30 to 50	APS
$Yb_2Si_2O_7/SiC$	100 to 150	SPPS
Yb_2SiO_5	100 to 150	SPPS

Both EBC systems were tested regarding their oxidation resistance at 1200 °C for 100 h in furnace air. Additionally, hot gas corrosion tests were conducted in a high-temperature burner rig at atmospheric pressure [11]. The coated test samples were blown directly by the hot gas in a reactor tube of solid-state sintered SiC with an inner diameter of 30 mm. The hot gas was composed of the combustion products of natural gas in air and additional water vapor. The conditions of the corrosion tests are summarized in Table 2. Further details regarding the burner rig test are described in [11].

Table 2. Burner rig test performed.

Temperature/°C	Flow Speed/m/s	Water Vapor Pressure/bar	Testing Time/h
1200	100	0.2	100

After both tests, the microstructure of the samples was characterized by means of polished cross-sections with field-emission scanning electron microscopy (Ultra 55, Zeiss, Oberkochen, Germany). Information about the composition of the different layers after oxidation and corrosion was obtained by using energy-dispersive X-ray spectroscopy (EDX; ISIS Si (Li) detector).

3. Results

A summary of the weight changes of the two coating systems in comparison to the base CMC substrate without coating observed during oxidation and the burner rig test is summarized in Table 3. A weight gain was observed for all materials investigated after both tests.

Table 3. Weight change after oxidation and burner rig test, 1200 °C, 100 h. Comparison of EBC-coated systems with base ceramic matrix composites (CMC) material without EBC coating.

Material/EBC	Oxidation		Burner Rig Test	
	m/g	m/m/g/g	m/g	m/m/g/g
CMC	0.027	0.009	0.094	0.046
Al_2O_3-YAG	0.048	0.016	0.19	0.064
Si-$Yb_2Si_2O_7$/SiC-Yb_2SiO_5	0.009	0.003	0.045	0.016

The values are the average of the three samples each. In connection with the interpretation of these results, some inaccuracies as a consequence of inhomogeneous EBC layers (thickness, pores, or cracks) should be considered. Furthermore, during the burner rig test, both processes, weight gain and weight loss, were observed. These results cannot be correlated directly with the TGO scale thickness obtained after the tests; however, some general tendencies can be followed as described in the microstructural discussion of the material investigated.

The main results were obtained by comparison of the microstructure of the CMC with different EBCs after oxidation and hot gas corrosion tests both at 1200 °C and 100 h. After oxidation, a weight gain was observed. During the test, oxygen diffused through the EBC layer and finally reacted at the first non-oxide surface in the system (base SiC_F-$SiC(N)$ composite or Si bond coat). Usually, a TGO layer was formed. In the case of the hot gas corrosion test, a second reaction has to be considered as well. Water vapor penetrated through the EBC layer, reacted with the SiO_2 of the TGO layer, and formed volatile $Si(OH)_4$. As a consequence of the high gas speed, the equilibrium of the corrosion reaction in Equation (1) was strongly shifted to the formation of $Si(OH)_4$. In this way, the water vapor corrosion of the TGO became more dominant, resulting in material loss and a gap formation between the silica TGO and the EBC top layer.

This behavior is described in Figure 1 showing the comparison of polished cross-sections of the microstructure after oxidation and corrosion of a model EBC system consisting of a Si bond coat and an Yb_2SiO_5 top coat. While a TGO was formed (A) after oxidation, corrosion processes with the formation of gaps at the Yb_2SiO_5-Si interface (B) were observed.

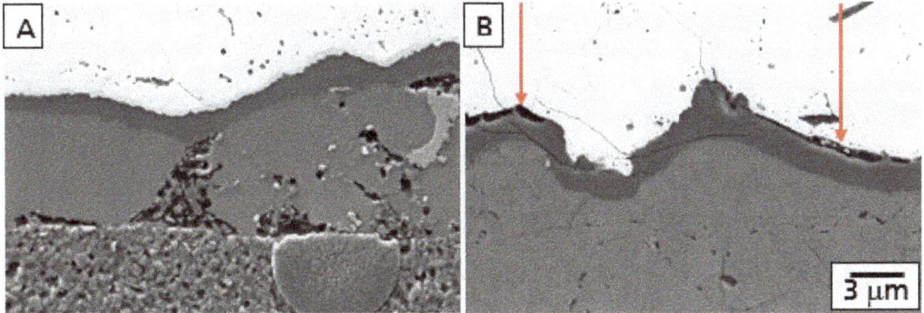

Figure 1. Polished cross-section of TGO in Si-Y_2SiO_5 EBC system after (**A**) 100 h oxidation at 1200 °C in air and (**B**) 100 h hot gas corrosion at 1200 °C.

3.1. CMC Substrate Material

The results of the high-temperature tests of the base CMC without EBC were characterized by investigation of the microstructure of the surface region. After both tests, oxidation and the burner rig, a weight gain was observed. During oxidation, a protective layer of mainly SiO_2 was formed at the surface of the material, limiting the diffusion of oxygen into the material (Figure 2A). This behavior was comparable to dense monolithic SiC. Both oxidation (weight gain) and corrosion (weight loss) were observed during the burner rig test. Caused by the water vapor corrosion of the protecting SiO_2 surface layer, oxygen and water vapor were able to diffuse deeper into the material and oxidized the matrix and the fibers too (Figure 2B). As a consequence of these oxidation processes inside of the material, a higher weight gain was found after the burner rig test.

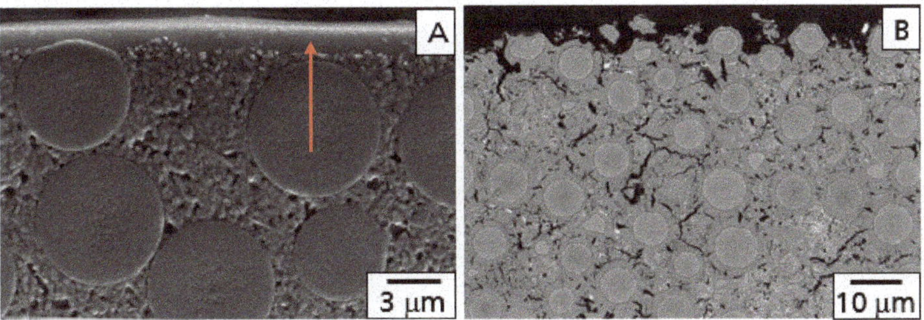

Figure 2. Microstructure of polished cross-sections of the base CMC after thermal treatment. (**A**) 100 h oxidation at 1200 °C; (**B**) 100 h burner rig test at 1200 °C.

3.2. Al_2O_3–YAG EBC Coating System

In the case of the material coated with Al_2O_3-YAG, a weight gain was observed after both tests. The values were found to be higher in comparison to the base CMC. Two effects are assumed to be the reason; first, a low protective ability caused by the inhomogeneity of the double layer with a high amount of cracks and porosity, and second, the TGO as the rate controlling factor for oxidation based on an alumosilicate glass with a significantly higher diffusion ability in comparison to the surface oxidation layer formed during oxidation of the base CMC [25].

The microstructure of the Al_2O_3-YAG EBC coating on the SiC_F-SiC(N) composite in the coated condition is illustrated in Figure 3.

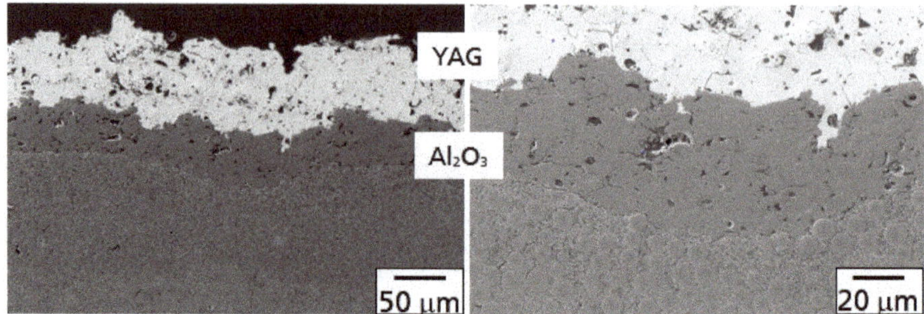

Figure 3. Microstructure of EBC consisting of Al$_2$O$_3$ bond coat and YAG top coat.

In principle, both processes, as described above, were observed after oxidation and the hot gas test. A comparison of polished cross-sections with the Al$_2$O$_3$/YAG coating is shown in Figure 4 after oxidation in air (A) and the burner rig test (B) at 1200 °C and 100 h.

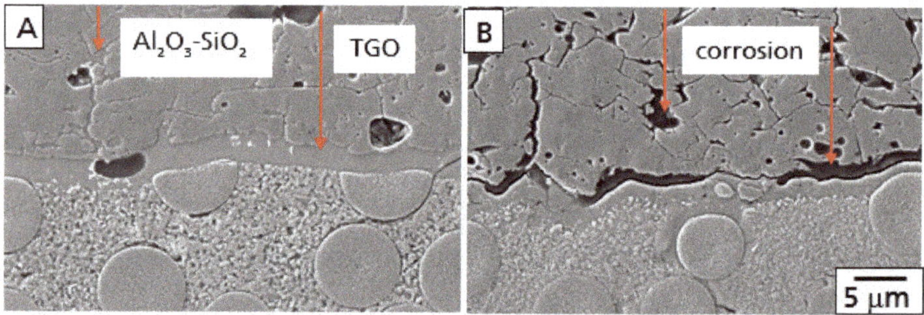

Figure 4. Comparison of microstructure in polished cross-sections of SiC$_F$/SiC(N) composite coated with Al$_2$O$_3$-YAG. (**A**) TGO formation and Al$_2$O$_3$-SiO$_2$ glass in and under the alumina layer after 100 h oxidation at 1200 °C. (**B**) Corrosion of TGO and grain boundary in the volume of Al$_2$O$_3$ layer after burner rig test 100 h and 1200 °C.

During the oxidation test, oxygen diffused through the YAG/Al$_2$O$_3$ layer and oxidized the SiC fibers and SiC(N)-matrix to SiO$_2$. Consequently, a TGO layer at the interface of the CMC base material and the Al$_2$O$_3$ bond coat was formed. Furthermore, a part of the oxidation product was found in the grain boundaries and triple junctions of the Al$_2$O$_3$ bond coat. The composition of the oxidation products in both the TGO and Al$_2$O$_3$ layer was a glassy alumosilicate (Figure 4A).

As described above, corrosion processes were observed additionally after the burner rig test (Figure 4B). The alumosilicate glass in both the TGO and the grain boundaries and triple junctions were found to be corroded, forming small pores and voids in the Al$_2$O$_3$ bond coat and gaps between the top of the TGO and the bond coat. With increasing time, both corrosion processes will damage the EBC system. Especially, the corrosion of the TGO will form large defects, finally leading to failure of the EBC. The smaller pores and voids in the alumina bond coat, however, are much more stable from a mechanical point of view.

Notwithstanding other damage mechanisms, this behavior opens an idea to the enhance high-temperature stability of the EBC in principle. By the shifting of the oxidation processes from the interface of the TGO into the volume of the bond coat, it should be possible to decelerate the TGO formation. Furthermore, the damage mechanism in the TGO (crystallization processes and corrosion) could be avoided, and only smaller corrosion defects in the bond coat should be observed.

3.3. Si–Yb$_2$Si$_2$O$_7$/SiC–Yb$_2$SiO$_5$ EBC Coating System

This mechanism described above was considered in the design of a three-layer coating system investigated next with a silicon bond coat, an intermediate layer consisting of Yb$_2$Si$_2$O$_7$ with SiC particles, and Yb$_2$SiO$_5$ as the top coat featured by a superior hot gas stability. The microstructure of this EBC is demonstrated in Figure 5. Few microcracks were observed in the Yb$_2$SiO$_5$ top layer, probably as a consequence of the CTE mismatch between the top and intermediate layer (CTE Yb$_2$Si$_2$O$_7$ 4.2 × 10^{-6} K^{-1}; CTE Yb$_2$SiO$_5$ 6.8 × 10^{-6} K^{-1}).

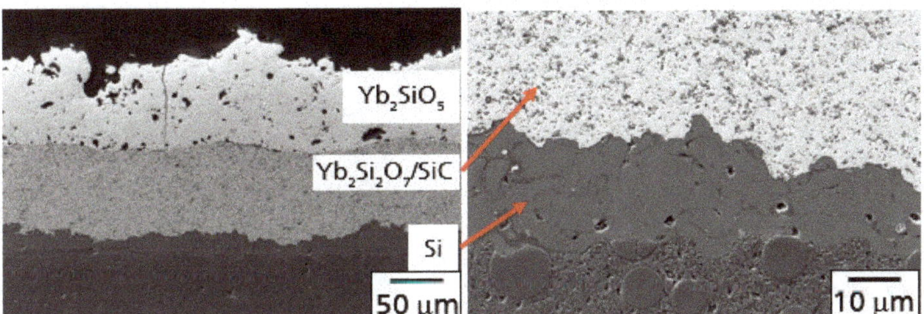

Figure 5. Polished cross-section of EBC layer system with Si bond coat, Yb$_2$Si$_2$O$_7$/SiC, and Yb$_2$SiO$_5$ layer fabricated by atmospheric plasma spaying (Si) and suspension plasma spraying (Yb$_2$Si$_2$O$_7$/SiC, Yb$_2$SiO$_5$).

In comparison to the other materials investigated, only a small weight gain was measured after both tests at elevated temperatures. This should be caused by a protecting function, especially of the Yb$_2$Si$_2$O$_7$/SiC layer. Oxygen diffusion through the EBC layers and the oxidation reaction in the Yb$_2$Si$_2$O$_7$/SiC layer were found to be the main processes observed after the oxidation test. The SiC particles in the intermediate layer were oxidized, consequently leading to the formation of a SiO$_2$ scale on the SiC particles. The diffusion of oxygen through this SiO$_2$-based layer controlled the oxidation process of the whole system during the testing time performed. A typical example for the mechanism is demonstrated in Figure 6A, presenting a polished cross-section at the interface of the Yb$_2$SiO$_5$ top coat and the Yb$_2$Si$_2$O$_7$/SiC intermediate layer.

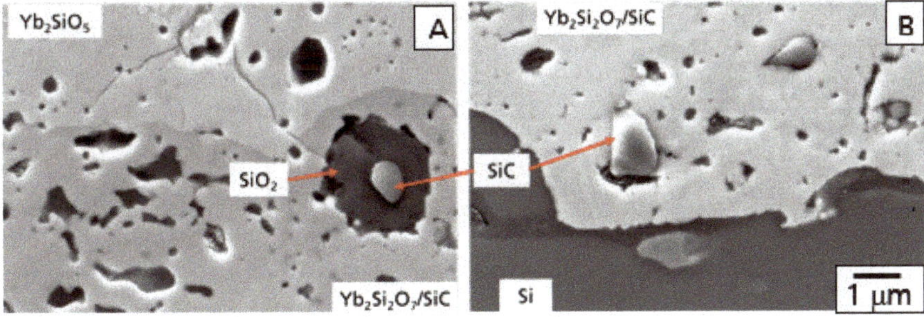

Figure 6. Polished cross-section of Si–Yb$_2$Si$_2$O$_7$/SiC–Yb$_2$SiO$_5$ EBC coating system. SiC particle in Yb$_2$Si$_2$O$_7$ layer at interfaces (**A**) Yb$_2$SiO$_5$–Yb$_2$Si$_2$O$_7$/SiC and (**B**) Yb$_2$Si$_2$O$_7$/SiC–Si after oxidation for 100 h and 1200 °C.

The SiC particles in this layer operated as a getter for the diffusion of oxygen, preventing further oxygen at diffusing deeper into the material. This effect is illustrated in Figure 6B with the microstructure at the interface between the Si bond coat and the $Yb_2Si_2O_7$/SiC intermediate layer. As the diffusion of oxygen reacted with SiC particles in the upper region of this layer, no oxidation processes were observed in this area. The SiC grains did not show an oxidation layer. Furthermore, the formation of a TGO was not observed. Similar results on the graded oxidation of SiC-particles have been reported in context of the mechanical self-healing ability of $Yb_2Si_2O_7$/SiC composites [26,27].

In Figure 7, the microstructure of this EBC system is shown after the 100 h hot gas corrosion test at 1200 °C. Although some additional cracks formed, the EBC coating is still intact, protecting the CMC from the direct hot gas corrosion attack (Figure 7A).

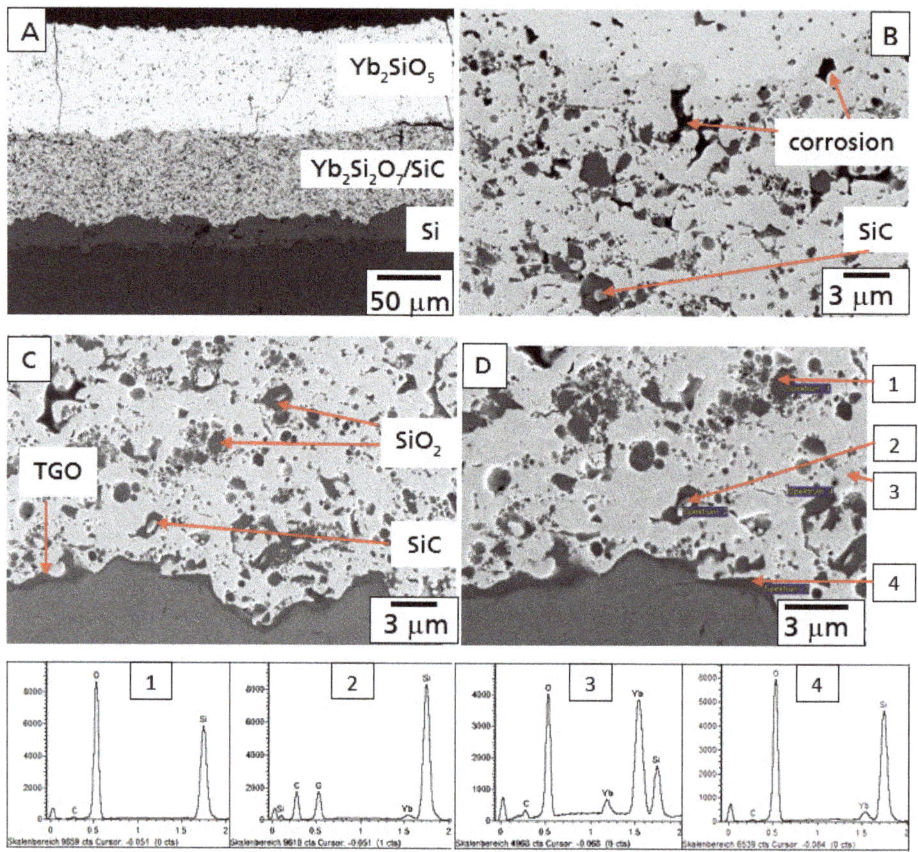

Figure 7. Polished cross-section of SiC$_F$/SiC(N) composite with three-layer EBC (Si–$Yb_2Si_2O_2$/SiC–Yb_2SiO_5) after 100 h hot gas test (**A**) at 1200 °C, overview; (**B**) at the interface to the Yb_2SiO_5 top coat and the $Yb_2Si_2O_7$/SiC interlayer, and (**C**) at the interface between the $Yb_2Si_2O_7$/SiC interlayer and the Si bond coat; energy-dispersive X-ray spectroscopy (EDX) analysis of different phases in the microstructure, analysis locations demonstrated in (**D**).

In comparison to the oxidation test, stronger oxidation processes were observed in the $Yb_2Si_2O_7$/SiC layer after the test in hot gas conditions. A higher amount of oxidants reached and oxidized the SiC particles in the intermediate layer of the EBC coating system as a consequence of additional water vapor permeation. In agreement to the literature [28,29], water vapor was found to be the dominant oxidant caused by the significantly higher solubility of H_2O in SiO_2. A second reasonable possibility is the formation of interconnected open splat boundaries in the Yb-silicate layer [26]. Further investigation has to be performed in this field. The SiC particles were found to be oxidized to SiO_2 up to the bottom of the $Yb_2Si_2O_7$/SiC layer. As a part of the oxidant reached the Si-bond coat the TGO layer started to grow (Figure 7C). Results of EDX analysis of microstructural details of this part are presented in Figure 7.

Additional processes were observed in the top region of the $Yb_2Si_2O_7$/SiC layer. Some of the SiO_2 areas formed by the oxidation of the SiC particles were found to be corroded by water vapor, finally leading to the formation of pores in the top region of the intermediate layer (Figure 7B).

4. Discussion

As already mentioned above, the penetration of oxidants into the EBC system and the following formation of a TGO cannot be prevented completely. However, with purposeful design of the EBC layer system, the transport processes and the following oxidation processes can be influenced. The first possibility should be the reduction of the transport processes for O_2 and H_2O. Coatings with a lower oxygen diffusion coefficient and water vapor solubility are reasonable trends for material development.

In the present study, the following oxidation and corrosion processes were modified. With the introduction of oxidable particles into an intermediate layer, it was possible to shift the oxidation processes from the interface between the Si bond coat and the Yb-silicate top coat to a volume in the intermediate layer. The processes occurring during oxidation and corrosion are schematically described in Figure 8.

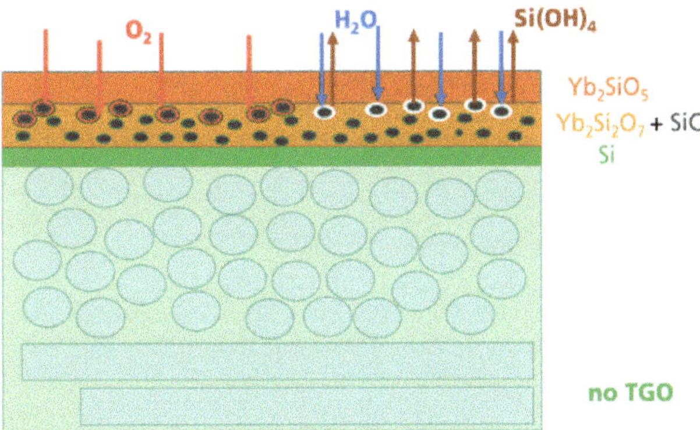

Figure 8. Schematic diagram of the oxidation and corrosion mechanism of SiC$_F$/SiC(N) composite with three-layer EBC system.

Independent of the thermal treatment, oxidation, or hot gas corrosion, the transport of oxidants and the following oxidation processes in the EBC system were found to be the first processes in the system. The penetrated O_2 and H_2O reacted with the SiC, resulting in the formation of a SiO_2-based shell surrounding the SiC particles. In this way, the SiC particles served as a getter for the further transport of oxidants into deeper regions of the whole system. As the oxidants could not reach the Si bond coat, the formation of the TGO was prevented. In a hot gas atmosphere with increased water vapor pressure and high gas velocity, corrosion processes were observed. The water vapor penetrated this region and reacted with the SiO_2 at the surface of the SiC particles and formed $Si(OH)_4$. A $Si(OH)_4$ gradient developed as a consequence of the high hot gas velocity outside, which was a driving force of the outward transport and evaporation of the $Si(OH)_4$. Small pores were found to be the result of the corrosion process.

The beneficial gettering function of the SiC particulate will be a temporary effect only. After longer oxidation time or at higher temperatures, the SiC particles will be consumed and oxygen will reach the silicon bond coat to form the TGO. However, with the incorporation of the SiC particles, to be oxidized during operation in hot gas atmosphere, the EBC system could be temporary stabilized. With the delayed formation of the TGO, their resulting damage mechanisms, cracking as a result of stresses by crystallization and phase transition processes and the gap formation caused by corrosion material loss, started at later application times. A very beneficial effect in terms of long-term stability and lifetime can only be achieved by simultaneous improvement of the oxygen and H_2O permeation behavior of the whole EBC layer system. The lower the transport of the oxidants into the material, the longer the gettering function of the SiC particles can be used. Further studies have to be performed to optimize the EBC system regarding composition and microstructure with special focus on the transport mechanisms during service in hot gas environments.

Author Contributions: H.K. conceptualized the idea; H.K., K.S. and W.K. developed the coating; K.S. and W.K. performed the experiments and analyzed data; H.K. supervised the project and acquired funding; H.K. prepared the manuscript; K.S. and W.K. contributed in editing and submission. All authors have read and agreed to the published version of the manuscript.

Funding: This research was funded by the German Federal Ministry for Education and Research, grant number 03EK3544C and Fraunhofer Funding MAVO CMC engine.

Acknowledgments: The authors gratefully thank B. Gronde and F.L. Toma for support in plasma spraying technology.

Conflicts of Interest: The authors declare no conflict of interest.

References

1. Endo, Y.; Tsuchiya, T.; Furuse, Y. Corrosion behavior of ceramics for gas turbines application, silicon based structural ceramics. *Ceram. Trans.* **1994**, *42*, 319–326.

2. Opila, E.J.; Hann, R.E. Paralinear Oxidation of CVD SiC in water vapor. *J. Am. Ceram. Soc.* **1997**, *80*, 197–205. [CrossRef]

3. Yuri, I.; Hisamatsu, T. Recession Rate Prediction for Ceramic Materials in Combustion Gas Flow. In Proceedings of the ASME Turbo Expo 2003, Atlanta, GA, USA, 16–19 June 2003.

4. Filsinger, D.; Schulz, A.; Wittig, S.; Taut, C.; Klemm, H.; Wötting, G. Model Combustor to Assess the Oxidation Behavior of Ceramic Materials under Real Engine Conditions. In Proceedings of the ASME Turbo Expo '99, Indianapolis, Indianapolis, IN, USA, 7–10 June 1999.

5. More, L.K.; Tortorelli, P.F.; Ferber, M.K.; Keiser, L.R. Observations of accelerated silicon carbide recession by oxidation at high water-vapor pressures. *J. Am. Ceram. Soc.* **2000**, *83*, 211–223. [CrossRef]

6. Klemm, H.; Schubert, C.; Taut, C.; Schulz, A.; Wötting, G. Corrosion of Non-Oxide Silicon-Based Ceramics in a Gas Turbine Environment. In Proceedings of the 7th Symposium on Ceramic Materials & Components for Engines, 2000, Weinheim, Germany, 29 March 2001; WILEY-VCH Verlag: Weinheim, Germany, 2001; pp. 153–156.

7. Lee, K.N.; Fritze, H.; Ogura, Y. Coatings for engineering ceramics. In *Progress in Ceramic Gas Turbine Development*; van Roode, M., Ferber, M., Richerson, D.W., Eds.; ASME Press: New York, NY, USA, 2003; Volume 2, pp. 641–664.

8. Lee, K.N. Environmental barrier coatings for CMCs. In *Ceramic Matrix Composites*; Bansal, N.P., Lamon, J., Eds.; Wiley: New York, NY, USA, 2015; pp. 430–451.

9. Lee, K.N. Current status of environmental barrier coatings for Si-based ceramics. *Surf. Coat. Technol.* **2000**, *133*, 1–7. [CrossRef]

10. Lee, K.N.; Fox, D.S.; Bansal, N.P. Rare earth silicate environmental barrier coatings for SiC/SiC composites and Si3N4 ceramics. *J. Eur. Ceram. Soc.* **2005**, *25*, 1705–1715. [CrossRef]

11. Fritsch, M. Heißgaskorrosion keramischer Werkstoffe in H2O-haltigen Rauchgasatmosphären. In *Dissertationsschrift, TU Dresden*; Fraunhofer IRB Verlag: Stuttgart, Germany, 2007; ISBN 978-3-8167-7588-1.

12. Klemm, H. Silicon nitride for high-temperature applications. *J. Am. Ceram. Soc.* **2010**, *93*, 1501–1522. [CrossRef]

13. Ueno, S.; Ohji, T.; Lin, H.T. Recession behavior of Yb2Si2O7 phase under high speed steam jet at high temperatures. *Corros. Sci.* **2008**, *50*, 178–182. [CrossRef]

14. Yuri, I.; Hisamatsu, T.; Ueno, S.; Ohji, T. Exposure test results of Lu2Si2O7 in combustion gas flow at high temperature and high speed. *Jpn. Soc. Mech. Eng.* **2001**, *44*, 520–527.

15. Hong, Z.; Cheng, L.; Zhang, L.; Wang, Y. Water vapor corrosion behavior of scandium silicates at 1400 °C. *J. Am. Ceram. Soc.* **2009**, *92*, 193–196. [CrossRef]

16. Richards, B.T.; Young, K.A.; de Francqueville, F.; Sehr, S.; Begley, M.R.; Wadley, H.N.G. Response of ytterbium disilicate–silicon environmental barrier coatings to thermal cycling in water vapor. *Acta. Mater.* **2016**, *106*, 1–14. [CrossRef]

17. Bakan, E.; Sohn, Y.J.; Kunz, W.; Klemm, H.; Vaßen, R. Effect of processing on high-velocity water vapor recession behavior of Yb-silicate environmental barrier coatings. *J. Europ. Ceram. Soc.* **2019**, *39*, 1507–1513. [CrossRef]

18. Lee, K.N.; Fox, D.S.; Eldridge, J.I.; Zhu, D.; Robinson, R.C.; Bansal, N.P.; Miller, R.A. Upper temperature limit of environmental barrier coatings based on mullite and BSAS. *J. Am. Ceram. Soc.* **2003**, *86*, 1299–1306. [CrossRef]

19. Lee, K.N. Yb2Si2O7 Environmental barrier coatings with reduced bond coat oxidation rates via chemical modifications for long life. *J. Am. Ceram. Soc.* **2019**, *102*, 1507–1521. [CrossRef]

20. Xu, Y.; Hu, X.; Xu, F.; Li, K. Rare earth silicate environmental barrier coatings: Present status and prospective. *Ceram. Int.* **2017**, *43*, 5847–5855. [CrossRef]

21. Zhong, X.; Niu, Y.; Li, H.; Zhou, H.; Dong, S.; Zheng, X.; Ding, C.; Sun, J. Thermal shock resistance of tri-layer Yb2SiO5/Yb2Si2O7/Si coating for SiC and SiC-matrix composites. *J. Am. Ceram. Soc.* **2018**, *101*, 4743–4752. [CrossRef]

22. Lu, Y.; Luo, L.; Liu, J.; Zhu, C.; Wang, Y. Failure mechanism associated with the thermally grown silica scale in environmental barrier coated C/SiC composites. *J. Am. Ceram. Soc.* **2016**, *99*, 2713–2719. [CrossRef]

23. Klemm, H.; Herrmann, M.; Schubert, C. High temperature oxidation and corrosion of silicon-based nonoxide ceramics. *J. Eng. Gas. Turbines Power* **2000**, *122*, 13–18. [CrossRef]

24. Schönfeld, K.; Klemm, H. Interaction of fiber matrix bonding in SiC/SiC ceramic matrix composites. *J. Europ. Ceram. Soc.* **2019**, *39*, 3557–3565. [CrossRef]

25. Courtright, E.L. Engineering property limitations of structural ceramics and ceramic composites above 1600 °C. *Ceram. Eng. Sci. Proc.* **1991**, *12*, 1725–1744.

26. Kunz, W. Entwicklung selbstheilender Materialien zum Korrosionsschutz für Keramiken in Hochtemperaturanwendungen. In *Dissertationsschrift, TU Dresden*; Fraunhofer: Munich, Germany, 2019; ISBN 978-3-8396-1507-2.

27. Kunz, W.; Klemm, H. Self-healing EBC material for gas turbine applications. Advances in high temperature ceramic matrix composites and materials for sustainable development. *Ceram. Trans.* **2018**, *263*, 173–185.

28. Deal, B.E.; Grove, A.S. General relationship for the thermal oxidation of silicon. *J. Appl. Phys.* **1965**, *36*, 3770–3778. [CrossRef]

29. Opila, E.J. Variation of the oxidation rate of silicon carbide with water-vapor pressure. *J. Am. Ceram. Soc.* **1999**, *82*, 625–636. [CrossRef]

Article

Structural Stabilization of Mullite Films Exposed to Oxygen Potential Gradients at High Temperatures

Satoshi Kitaoka *, Tsuneaki Matsudaira, Naoki Kawashima, Daisaku Yokoe, Takeharu Kato and Masasuke Takata

Japan Fine Ceramics Center (JFCC), Nagoya 456-8587, Japan; matsudaira@jfcc.or.jp (T.M.); kawashima@jfcc.or.jp (N.K.); yokoe@jfcc.or.jp (D.Y.); tkato@jfcc.or.jp (T.K.); m_takata@jfcc.or.jp (M.T.)
* Correspondence: kitaoka@jfcc.or.jp; Tel.: +81-52-871-3500

Received: 23 August 2019; Accepted: 23 September 2019; Published: 30 September 2019

Abstract: The oxygen shielding properties of polycrystalline $Al_{4+2x}Si_{2-2x}O_{10-x}$ (mullite) films applied as environmental barrier coatings (EBCs) on SiC fiber-reinforced SiC matrix composites (SiC/SiC) are determined by the grain boundary (GB) diffusion of oxide ions in the films, from the higher oxygen partial pressure (P_{O_2}) surface to the lower P_{O_2} surface, with simultaneous GB diffusion of Al ions in the opposite direction. Herein, strategies to improve the oxygen shielding and phase stability of these films when applied to SiC/SiC substrates through bond coats are proposed, based on oxygen permeation data for mullite at high temperatures. The validity of these strategies is verified using experimental trials at 1673 K with bilayer specimens consisting of mullite films and bond coat substrates, serving as model EBCs. The data show that employing a bond coat made of β′-SiAlON rather than Si provides a source of Al for the overlying mullite film that greatly improves the phase stability of the film in the vicinity of the junction interface. Because the minimum equilibrium P_{O_2} values required to form SiO_2 due to oxidation of the β′-SiAlON on a thermodynamic basis are significantly larger than those for oxidation of Si, the inward GB diffusion of oxide ions is effectively retarded, resulting in excellent oxygen shielding characteristics.

Keywords: EBCs; mullite; SiAlON; diffusion; grain boundary; oxygen permeability

---

1. Introduction

Environmental barrier coatings (EBCs) can play a key role in enabling SiC fiber-reinforced SiC matrix (SiC/SiC) composites to be employed as advanced hot-section components in airplane engines, as a means of realizing exceptional fuel efficiency. EBCs must exhibit excellent oxygen/water vapor shielding characteristics and thermomechanical durability in severe combustion environments [1–7]. Thus, a multilayer structure is generally adopted, with each layer having its own unique properties. Naturally, EBCs are exposed to a large oxygen potential gradient ($d\mu_O$) at elevated temperatures. This gradient results in the inward diffusion of oxide ions and outward diffusion of cations, according to the Gibbs-Duhem relationship. Cation transport through an oxide layer typically induces decomposition of the oxide, and can thus cause the EBC structure to collapse. Therefore, the development of a robust EBC with excellent gas shielding characteristics requires both a good understanding of, and control over, mass transfer within the EBC. However, even though the performance of an EBC is greatly affected by both its constituent materials and its microstructure depending on process conditions, there is a crucial lack of knowledge regarding the environmental shielding properties of these materials. Consequently, studies intended to improve EBC performance tend to be based on trial-and-error.

In prior work, our group elucidated the mass transfer processes within the constituent oxides of EBCs by determining the oxygen permeability values of wafers cut from sintered bodies serving as model EBCs [8–13]. This technique is believed to accurately evaluate the environmental shielding characteristics of EBC materials under steady state conditions. This is possible because both the $d\mu_O$ applied to the wafers and the diffusion length are kept constant. In this technique, the upper and lower surfaces of the wafer are exposed to gases with different oxygen partial pressures (P_{O_2}) such that the dissociative adsorption of oxygen molecules will proceed at the surface having the higher P_{O_2} (the P_{O_2}(hi) surface), while the reverse reaction progresses on the opposite surface having a lower P_{O_2} (the P_{O_2}(lo) surface). Therefore, the mass transfer mechanisms through the oxides comprising the EBC can be monitored in a facile manner.

Mullite, which is often given the formula $Al_{4+2x}Si_{2-2x}O_{10-x}$, is regarded as a candidate material for multilayered EBCs because its thermal expansion coefficient is close to that of SiC and it is chemically compatible with Si-based ceramics [5]. The oxygen shielding properties of polycrystalline mullite, in which $x = 0.25$ (corresponding to $3Al_2O_3 \cdot 2SiO_2$ and referred to as so-called "stoichiometric" or stable mullite [14]) are evidently superior to those of $Yb_2Si_2O_7$ and Yb_2SiO_5 [10–12], although the resistance of mullite to volatilization by water vapor is far inferior to that of the Yb-silicates [15]. Oxygen permeation through these complex oxides is controlled by the inward diffusion of oxide ions and the outward diffusion of cations (M), such as Al ions in the case of mullite and Yb ions in the case of Yb-silicates, through grain boundaries (GBs) [10–12]. The mobility of Si ions, which comprise the majority of the complex oxides, is evidently too low to contribute to oxygen permeation. Thus, going from the SiC/SiC substrate to the top coat, an EBC typically consists of a Si-based non-oxide bond coat, a mullite layer and Yb-silicate layers. This configuration improves both the oxygen shielding properties and water vapor volatilization resistance of the overall EBC system.

Using the oxygen permeation data resulting from prior work, the oxygen permeability constants of each oxide layer constituting a multi-layered EBC, as well as the contributions of oxide ions and cations to oxygen permeation through the oxide, mass transfer parameters such as chemical potentials, GB diffusion coefficients and fluxes of both ions, can all be calculated at arbitrary temperatures and $d\mu_O$ values [8–12]. The ability to determine mass transfer through oxides provides useful structural information regarding layer arrangement, layer thickness and grain size that can assist in designing EBCs with excellent oxygen shielding characteristics together with phase stability.

When a mullite layer with superior oxygen shielding characteristics is applied to a conventional bond coat made of Si, the mullite is exposed to an extremely large $d\mu_O$ as part of the overall EBC system. The subsequent mass transfer of species diffusing through the mullite is expected to greatly affect the structural stability of the EBC. The present study assessed strategies to improve the oxygen shielding properties and phase stability of a mullite film applied to a SiC/SiC substrate through a Si-based non-oxide bond coat, using previously acquired oxygen permeation data [10] for mullite at high temperatures. The validity of these strategies was verified via experimental work at 1673 K using bilayer specimens consisting of a mullite film and a bond coat substrate, serving as model EBCs.

2. Design to Improve the Oxygen Shielding and Phase Stability of Mullite

In this study, the oxygen shielding obtained from a mullite film formed on a bond coat substrate was examined, employing a simplified EBC system. As described in the Introduction, the oxidation of a Si-based bond coat proceeds via the inward GB diffusion of oxide ions and the outward GB diffusion of Al ions through the exposed upper mullite film following the application of a large $d\mu_O$ at high temperatures. The transport of Al ions reduces the phase stability of the mullite in the vicinity of the junction interface between the mullite film and the underlying bond coat. Therefore, it would be beneficial to design a bond coat capable of supplying Al ions to the adjacent region in the upper mullite film that is otherwise depleted of Al.

β'-SiAlON, having the formula $Si_{6-z}Al_zO_zN_{8-z}$, is a particularly promising candidate for bond coat applications because it provides superior heat resistance compared to Si and its thermal expansion coefficient is close to that of SiC/SiC [16]. Moreover, β'-SiAlON can dissolve significant quantities of Al, and so could serve to supply Al to the mullite film. In addition, even though the concentration of Al in the β'-SiAlON would be decreased during this process, the β'-SiAlON is able to transition to other oxynitrides during this process, such as O′-SiAlON and the X phase. This continuous variation in phase structure means that the β′-SiAlON is less likely to delay formation of SiO_2 and subsequent crystallization such as cristobalite [17,18]. It is well known that phase transition of cristobalite from β-type to α-type at approximately 540 K during cooling leads to significant shrinkage (3.9 vol.%) [19], resulting in initiation of microcracks at the interfaces and/or in the cristobalite layer. Furthermore, in the case of Si_2N_2O that is a unit structure of O′-SiAlON having the formula $Si_{2-x}Al_xO_{1+x}N_2$, where the x-value varies from zero to ~0.2, the thermodynamic equilibrium P_{O_2} values, in which Si_2N_2O and SiO_2 coexist, are larger than those for Si and SiC [20,21]. Therefore, it is expected that the level of oxygen shielding required for oxide films to protect oxidation of underlying SiAlON-based ceramics can be lower than that for Si and SiC. Thus, incorporating a SiAlON bond coat is advantageous for maintaining long-term adhesion between the mullite layer and the bond coat. In the work reported herein, the effects of the bond coat material on the oxygen shielding and phase stability of the mullite film are discussed, based on comparing β′-SiAlON and Si bond coats.

When a mullite film is exposed to a $d\mu_O$ at a high temperature, such that each surface of the film is subjected to a different P_{O_2}, and P_{O_2}(hi) >> P_{O_2}(lo), the oxygen permeability constant, PL, per unit GB length can be expressed as Equation (1) [10]:

$$\frac{PL}{S_{gb}} = \frac{A_{Al}}{4S_{gb}}\left(P_{O_2}(\text{hi})^{3/16} - P_{O_2}(\text{lo})^{3/16}\right) + \frac{A_O}{4S_{gb}}\left(P_{O_2}(\text{hi})^{-1/4} - P_{O_2}(\text{lo})^{-1/4}\right) \tag{1}$$

where P is the oxygen permeability and L is the wafer thickness. The oxygen permeability term, PL, normalized by the GB density, S_{gb}, is dependent only on the GB characteristics and is unaffected by the grain size of the polycrystalline mullite. The experimental constants A_{Al} and A_O are related to the adsorption reactions of oxygen molecules that occur on the P_{O_2}(hi) surface and subsequent movement of Al ions and oxide ions during the oxygen permeation. A_{Al} and A_O normalized by S_{gb} are given by Equations (2) and (3), respectively [10]:

$$\frac{|A_{Al}|}{S_{gb}} = 1.139 \times 10^{-10} \exp\left(\frac{-235\ \text{kJ/mol}}{RT}\right) \tag{2}$$

and

$$\frac{|A_O|}{S_{gb}} = 3.612 \times 10^{-3} \exp\left(\frac{-518\ \text{kJ/mol}}{RT}\right) \tag{3}$$

where R is the gas constant, and T is the absolute temperature. Thus, the former and latter terms enclosed in parentheses in Equation (1) are related to the movement of Al ions and oxide ions, respectively. In addition, the P_{O_2}(lo) term in Equation (1) is assumed to equal the minimum equilibrium partial pressure ($P_{O_2,eq}$) required to form SiO_2 via the oxidation of Si or β′-$Si_3Al_3O_3N_5$. The $P_{O_2,eq}$ values associated with this work were calculated on a thermodynamic basis using the FactSage free energy minimization computer code in conjunction with various databases, including FactPS, FToxide and FTOxCN. In the case of β′-$Si_3Al_3O_3N_5$, α-Al_2O_3 is formed by the partial oxidation of β′-SiAlON at an equilibrium P_{O_2} value below $P_{O_2,eq}$. The ongoing oxidation process eventually degrades the various oxynitride products such as O′-SiAlON and the X phase until only SiO_2 and mullite remain. In this study, the $P_{O_2,eq}$ required to form SiO_2, which would be expected to seriously damage the adhesion of the mullite film, was examined on this basis.

Figure 1 plots the $P_{O_2,eq}$ values for the oxidation of the bond coat materials Si and β′-$Si_3Al_3O_3N_5$ as functions of temperature. The line for Si in Figure 1 corresponds to the oxygen equilibrium partial pressure at the phase boundary between Si and SiO_2 and is in agreement with that in the Si-O volatility diagram [20]. In the case of SiAlON, the upper side corresponds to the stable region of SiO_2, while the lower side is equivalent to the region where any SiAlONs such as the β′, O′, and X phases exist. It is evident that the $P_{O_2,eq}$ values for both materials increase with increasing temperature, and that the values for the β′-$Si_3Al_3O_3N_5$ are larger than those for Si over the entire temperature range. This indicates that applying β'-$Si_3Al_3O_3N_5$ can lower the oxygen shielding performance required for the upper oxide film compared to Si. The oxygen permeability constants of mullite exposed to various $d\mu_O$ were calculated using Equations (1)–(3) with the $P_{O_2,eq}$ values in Figure 1.

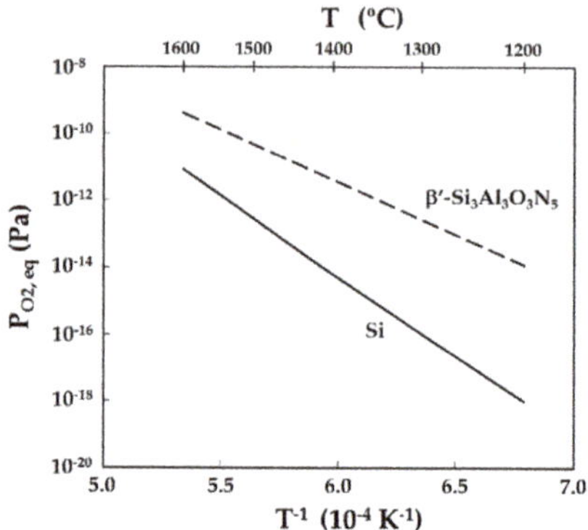

Figure 1. Temperature dependence of the minimum oxygen equilibrium partial pressures ($P_{O_2,eq}$) required to generate SiO_2 thermodynamically via the oxidation of the bond coat materials Si and SiAlON.

In the case of a mullite film on a Si bond coat (Figure 2a) with the mullite surface exposed to a P_{O_2}(hi) of 10^5 Pa, the plot showing the oxygen permeability constants related to the diffusions of both oxide ions and Al ions almost coincides with the line solely for the diffusion of oxide ions. Therefore, the oxygen permeation along the GBs in the mullite film is controlled by the inward GB diffusion of oxide ions over the entire temperature range. Using the same bilayer structure but a lower P_{O_2}(hi) of 1 Pa produced the data shown in Figure 2b, for which the total oxygen permeability constants and the contribution of the diffusion of oxide ions are almost equal to those in Figure 2a. However, the line associated with the diffusion of Al ions in Figure 2b drops to about one-tenth of that in Figure 2a. When the value of the P_{O_2}(hi) term that appears in Equation (1) was reduced to less than 10^2–10^3 times the $P_{O_2,eq}$, there was an evident decrease in the line for the diffusion of oxide ions. Thus, coating the mullite layer with a film made of an oxide such as an Yb-silicate [11,12], which provides inferior oxygen shielding, improves the phase stability of the mullite film but does not increase the overall oxygen shielding characteristics of the EBC. In the case of the mullite film on the β'-$Si_3Al_3O_3N_5$ bond coat with a P_{O_2}(hi) of 10^5 Pa, as shown in Figure 2c, the oxygen permeability constant is clearly decreased as a result of the significantly lower contribution of the diffusion of oxide ions. Specifically, the oxygen permeability constant due to the inward diffusion of oxide ions at 1673 K in Figure 2c was about one-fifth of the values for Figure 2a,b. It should be noted also that the contribution of the diffusion of Al ions remains much the same as in Figure 2a. The increase in the $P_{O_2,eq}$ values improves the oxygen shielding properties but not the phase stability of the mullite film.

Together, these results demonstrate that the oxygen permeation and the contribution of the diffusion of each ion through the mullite film are both significantly affected by the $d\mu_O$ value. This occurs because the GB diffusion coefficients of Al ions and oxide ions change greatly as the oxygen chemical potential (μ_O) in the thickness direction of the mullite film varies. That is, the GB diffusion coefficients of Al ions are much larger than those of oxide ions in regions associated with high μ_O values close to the P_{O_2}(hi) surface, while the reverse is true in regions with low μ_O values close to the P_{O_2}(lo) surface [10]. Therefore, an EBC structure comprising a sandwich-type configuration in which the mullite film is between a SiAlON bond coat and an oxide film providing a moderate degree of oxygen shielding (such as an Yb-silicate [11,12]) is expected to be especially effective. This structure should simultaneously improve the oxygen shielding properties and phase stability of the mullite film.

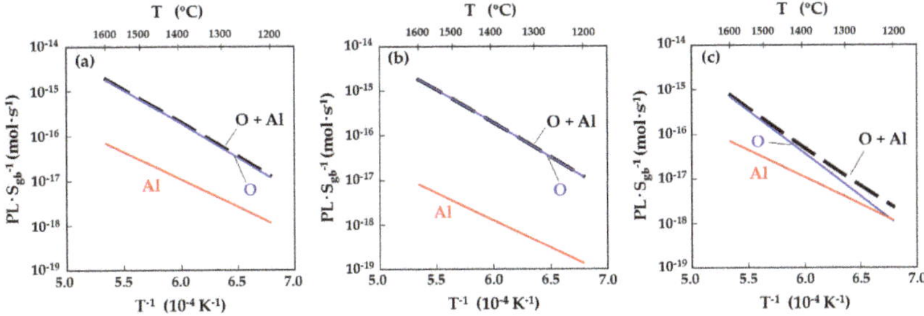

Figure 2. Temperature dependence of the oxygen permeability constants of mullite exposed to various oxygen potential gradients; (**a**) $P_{O_2}(\text{hi}) = 10^5$ Pa, $P_{O_2}(\text{lo}) = P_{O_2,\text{eq}}$ for the oxidation of Si, (**b**) $P_{O_2}(\text{hi}) = 1$ Pa, $P_{O_2}(\text{lo}) = P_{O_2,\text{eq}}$ for the oxidation of Si, and (**c**) $P_{O_2}(\text{hi}) = 10^5$ Pa, $P_{O_2}(\text{lo}) = P_{O_2,\text{eq}}$ for the oxidation of SiAlON.

3. Experimental Procedure

Figure 3 shows a schematic diagram of the procedure used to manufacture test specimens combining an oxygen shielding layer and an underlying bond layer made of Si or β'-SiAlON. In this fabrication process, raw commercially-available mullite powder (KM-101, KCM Corp., Nagoya, Japan) was molded in a uniaxial press at 20 MPa and then subjected to cold isostatic pressing at 250 MPa. The resulting green compacts were sintered in air at ambient pressure in two steps: 1573 K for 50 h followed by 2023 K for 5 h. High-resolution transmission electron microscopy (HR-TEM, TOPCON EM-002BF, Topcon Technohouse, Tokyo, Japan) indicated that there were no other phases at the GBs, meaning that there was direct contact between mullite grains. Raw β'-SiAlON powder having a z value of 2.94 (BSI3-001B, AG Material Inc., Taiwan) was hot pressed in a uniaxial press at 40 MPa under N$_2$ at a pressure of 0.7 MPa at 2023 K for 2 h. The relative densities of both the mullite and β'-SiAlON samples were greater than 99%. Parts with dimensions corresponding to the sections labeled types I, II and III in Figure 3, were then cut from the sintered samples and from a Si wafer (Shin-Etsu Chemical Corp. Ltd., Tokyo, Japan), and the joining surfaces of these samples were polished to a mirror-like finish. These parts were subsequently joined together using a uniaxial press at 50 MPa under reduced pressure at 1873 K for 1 h. Following this, one exposed mullite layer was ground and then polished to a final thickness of 0.25 mm. Finally, the specimens were heated in O$_2$ ($P_{O_2} = 10^5$ Pa) at 1673 K for 10 h. The microstructures of cross-sections of these samples in the vicinity of the bonding interfaces before and after the heat treatment were examined by scanning electron microscopy (SEM, Hitachi SU8000, Hitachi High-Technologies, Hitachinaka, Japan) and scanning transmission electron microscopy (STEM, TOPCON EM-002BF, Topcon Technohouse, Tokyo, Japan) with energy dispersive X-ray spectroscopy (EDS).

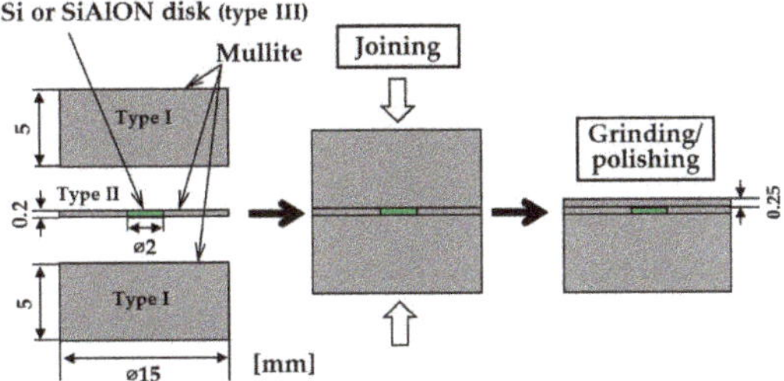

Figure 3. Schematic diagram of the procedure for manufacturing model samples corresponding to combinations of an oxygen shielding layer and an underlying bond layer made of Si or SiAlON.

4. Results and Discussion

The sample composed of a mullite film on a Si disk showed complete bonding of the mullite to the underlying Si disk in the thickness direction prior to heating, as shown in Figure 4a. No damage was observed in the vicinity of the joining interfaces in bright-field STEM images or in the corresponding EDS elemental maps of the cross-section around the interface. In contrast, after the heat treatment, damage to the mullite film near the interface was evident (Figure 4b). It is noteworthy that there was no degradation to the thicker 5 mm mullite layer at the opposite interface, which retained the same microstructure as before the heat treatment. Thus, the observed damage to the mullite is attributed to the increased driving force for the GB diffusion of Al and oxide ions resulting from the reduced film thickness. The uneven thickness of the damaged region in Figure 4b is likely related to variations in the GB density of the mullite film adjacent to the interface, because the GBs in the mullite provide paths for the rapid diffusion of ions.

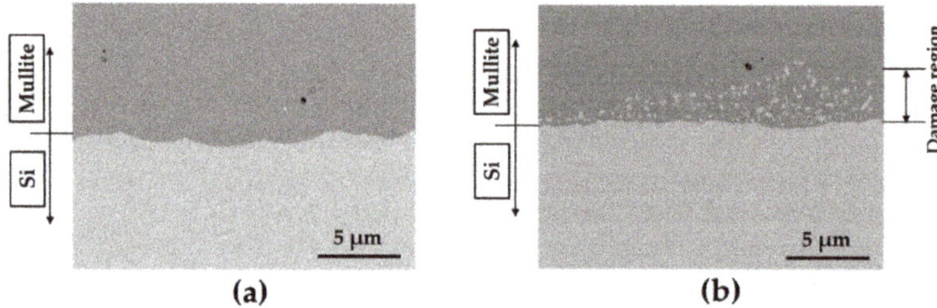

Figure 4. Scanning electron microscopy (SEM) images of cross-sections of the bonding interfaces of samples with a 0.25 mm thick mullite layer on an underlying Si disk (**a**) before and (**b**) after annealing in oxygen at 1673 K for 10 h.

Figure 5 presents a bright-field STEM image and the corresponding EDS elemental maps of the cross-section around the bonding interface of the annealed sample shown in Figure 4b. Because Al ions diffuse along the GBs in the mullite film in response to the application of the $d\mu_O$, both SiO_2-rich mullite and SiO_2 are expected to be formed by the decomposition of mullite near the interface. However, the damaged region shown in Figure 4b instead was found to consist of Al_2O_3 and Si. A Si-O phase was present only in areas enclosed by the mullite grains and located less than one micrometer from the interface. In addition, microstructural changes were not observed in the Si disk near the interface.

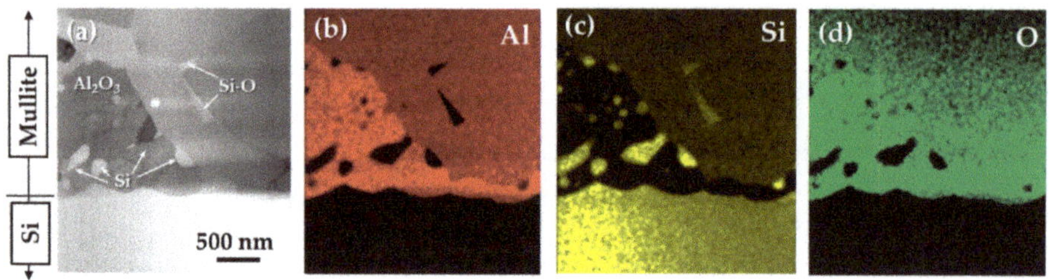

Figure 5. Bright-field scanning transmission electron microscopy (STEM) image and the corresponding energy dispersive X-ray spectroscopy (EDS) elemental maps of a cross-section of the bonding interface of the annealed sample shown in Figure 4b: (**a**) The bright-field image and EDS maps for (**b**) Al, (**c**) Si and (**d**) O.

Figure 6 presents a plot of the free energy curve for mullite in a SiO_2-Al_2O_3 system at 1673 K. Based on this plot, the concentration of Al_2O_3 in the mullite near the interface should change from an initial value, x_i, to a new value on the SiO_2-rich side of the plot, as a result of the outward GB diffusion of Al ions. When the Al_2O_3 concentration reaches a critical value, x_{cr}, a portion of the mullite will break down to produce both SiO_2-rich mullite and SiO_2, corresponding to the Si–O phase seen in Figure 5. In this study, the specimens were fabricated by joining parts made of mullite and Si using hot pressing with a carbon heater, and so the μ_O for oxygen dissolved in the Si disk was likely extremely low. Consequently, the SiO_2 formed at GBs of the mullite layer in the vicinity of the bonding interface due to the outward GB diffusion of Al ions was considered to be easily reduced to produce Si particles and oxygen that immediately dissolved in the Si substrate. Thus, voids under negative pressure were thought to be formed adjacent to the Si particles derived from the mullite layer. When mullite is exposed to a total pressure of 10^{-2} Pa or less at 1673 K, the decomposition reaction to form Al_2O_3 and SiO gas proceeds thermodynamically. Therefore, the mullite located around the voids was probably decomposed, resulting in generation of the Al_2O_3 particles. The voids might be also filled with Si that migrated from the Si disk, since the exposure temperature was near the melting point (1687 K) of Si. On the other hand, although the amount of oxygen permeated through the mullite layer calculated from Equations (1)–(3) was much larger than the value corresponding to the solid solubility limit of oxygen in Si [22], obvious SiO_2 scale formation due to oxidation of the Si disk was not observed at the interface. Oxide ions diffused inward along the GBs in the mullite layer might have been captured by oxygen vacancies in the mullite layer before reaching the interface between the mullite layer and the Si disk, in which a large amount of the vacancies were created around the GBs in the mullite layer during sample preparation in the reducing atmosphere.

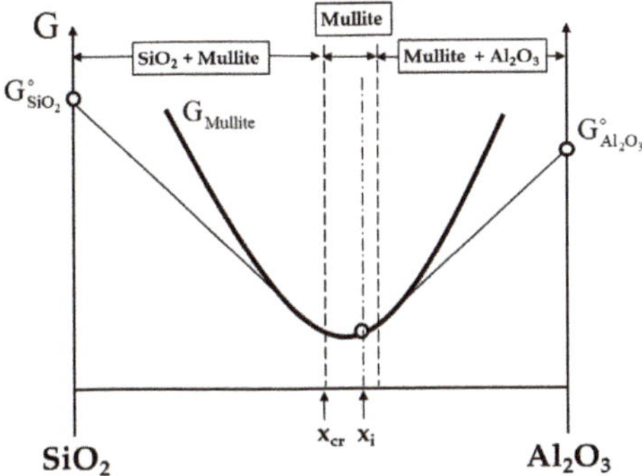

Figure 6. Schematic diagram of the free energy curve for mullite in a SiO_2-Al_2O_3 system at 1673 K. x_i and x_{cr} are the concentrations of alumina in mullite before annealing and in the mullite decomposition stage of annealing, respectively.

In the case of an actual EBC system in which a mullite layer is formed on a Si bond coat using an atmospheric plasma spray process, some ambient oxygen will dissolve in the lower Si bond coat during the high temperature coating, resulting in a sufficiently high μ_O in both the bond coat and mullite film compared with the specimens employed in this study. Therefore, a large amount of Si–O phase will be formed by decomposition of the mullite film due to the outward diffusion of Al ions and also by the oxidation of Si due to the inward diffusion of oxide ions, and will be present near the interface in the absence of oxide reduction.

Figure 7 provides TEM images of cross-sections around the bonding interface of a sample in which a 0.25 mm thick mullite layer was applied overtop of a SiAlON disk, before and after annealing in oxygen at 1673 K for 10 h. Figure 8 shows a bright-field STEM image and the corresponding EDS elemental maps of a cross-section around the bonding interface of the annealed sample shown in Figure 7b. The microstructures around the interfaces before and after the heat treatment are seen to be similar, and there is no damage to the mullite film resulting from the interdiffusion of Al and oxide ions. This result is believed to be related to the reduced inward GB diffusion of oxide ions due to the increased μ_O value in the vicinity of the interface and to the improved phase stability of the mullite film due to the Al supply from the underlying SiAlON.

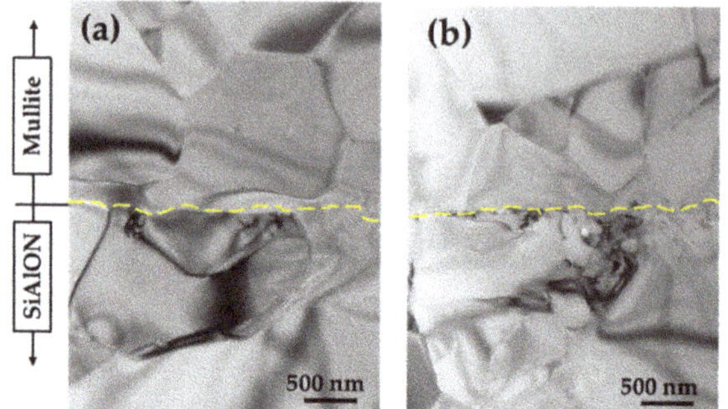

Figure 7. Transmission electron microscopy (TEM) images of a cross-section of the bonding interfaces of a sample with a 0.25 mm thick mullite layer on an underlying SiAlON disk (**a**) before and (**b**) after annealing in oxygen at 1673 K for 10 h.

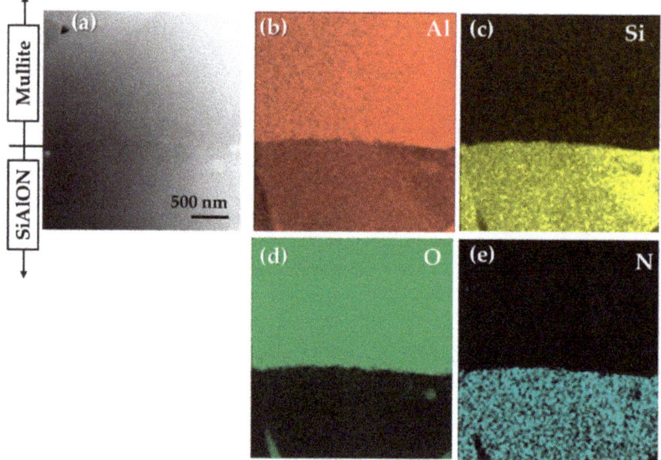

Figure 8. Bright-field STEM image and the corresponding EDS elemental maps of a cross-section of the bonding interface of the annealed sample shown in Figure 7b: (**a**) The bright-field image and EDS maps for (**b**) Al, (**c**) Si, (**d**) O and (**e**) N.

5. Conclusions

Strategies to improve the oxygen shielding properties and phase stability of mullite films on Si-based bond coats were examined and validity of these strategies was established using test models. When employing β′-SiAlON as the bond coat, the inward GB diffusion of oxide ions through the mullite film was suppressed due to an increase in the μ_O value at the interface between the mullite and the bond coat, resulting in improved oxygen shielding. Decomposition of the mullite film as a consequence of the outward diffusion of Al ions was retarded because the SiAlON served as a source of Al. Furthermore, the phase stability of the mullite was improved because the diffusion of Al through GBs was slowed due to the application of oxide films having suitable oxygen shielding characteristics.

Author Contributions: Conceptualization, S.K.; Methodology, S.K., N.K. and T.M.; Validation, S.K. and T.M.; Analysis, D.Y. and T.K., Investigation, S.K.; Writing—Original Draft Preparation, S.K.; Writing—Review and

Editing, S.K.; Supervision, S.K., M.T. and J.P.; Project administration, M.T. and J.P.; Funding acquisition, M.T. and J.P.

Funding: This work was partly supported by the Council for Science, Technology and Innovation (CSTI), the Cross-ministerial Strategic Innovation Promotion Program (SIP) "Structural Materials for Innovation" (Funding agency: JST, Japan Science and Technology Agency) and JSPS KAKENHI Grant Number JP19H05785.

Conflicts of Interest: The authors declare no conflict of interest.

References

1. Lee, K.N.; van Roode, M. Environmental barrier coatings enhance performance of SiC/SiC ceramic matrix composites. *Am. Ceram. Soc. Bull.* **2019**, *98*, 46–53.
2. Poerschke, D.L.; Sluytman, J.S.V.; Wong, K.B.; Levi, C.G. Thermochemical compatibility of ytterbia-(hafnia/silica) multilayers for environmental barrier coatings. *Acta Mater.* **2013**, *61*, 6743–6755. [CrossRef]
3. Cojocaru, C.V.; Lévesque, D.; Moreau, C.; Lima, R.S. Performance of thermally sprayed Si/mullite/BSAS environmental barrier coatings exposed to thermal cycling in water vapor environment. *Surf. Coat. Technol.* **2013**, *216*, 215–223. [CrossRef]
4. Xu, J.; Sarin, V.K.; Dixit, S.; Basu, S.N. Stability of interfaces in hybrid EBC/TBC coatings for Si-based ceramics in corrosive environments. *Int. J. Refract. Met. Hard Mater.* **2015**, *49*, 339–349. [CrossRef]
5. Richards, B.T.; Wadley, H.N.G. Plasma spray deposition of tri-layer environmental barrier coatings. *J. Eur. Ceram. Soc.* **2014**, *34*, 3069–3083. [CrossRef]
6. Richards, B.T.; Begley, M.R.; Wadley, H.N.G. Mechanisms of ytterbium monosilicate/mullite/silicon coating failure during thermal cycling in water vapor. *J. Am. Ceram. Soc.* **2015**, *98*, 4066–4075. [CrossRef]
7. Richards, B.T.; Sehr, S.; Francqueville, F.; Begley, M.R.; Wadley, H.N.G. Fracture mechanisms of ytterbium monosilicate environmental barrier coatings during cyclic thermal exposure. *Acta Mater.* **2016**, *103*, 448–460. [CrossRef]
8. Kitaoka, S.; Matsudaira, T.; Wada, M.; Saito, T.; Tanaka, M.; Kagawa, Y. Control of oxygen permeability in alumina under oxygen potential gradients at high temperature by dopant configurations. *J. Am. Ceram. Soc.* **2014**, *97*, 2314–2322. [CrossRef]
9. Ogawa, T.; Kuwabara, A.; Fisher, C.A.J.; Moriwake, H.; Matsunaga, K.; Tsuruta, K.; Kitaoka, S. A density functional study of vacancy formation in grain boundaries of undoped-alumina. *Acta Mater.* **2014**, *69*, 365–371. [CrossRef]
10. Kitaoka, S.; Matsudaira, T.; Yokoe, D.; Kato, T.; Takata, M. Oxygen permeation mechanism in polycrystalline mullite at high temperatures. *J. Am. Ceram. Soc.* **2017**, *100*, 3217–3226. [CrossRef]
11. Wada, M.; Matsudaira, T.; Kawashima, N.; Kitaoka, S.; Yokoe, D.; Takata, M. Mass transfer in polycrystalline ytterbium disilicate under oxygen potential gradients at high temperatures. *Acta Mater.* **2017**, *135*, 372–378. [CrossRef]
12. Kitaoka, S.; Matsudaira, T.; Wada, M.; Yokoi, T.; Yamaguchi, N.; Kawashima, N.; Ogawa, T.; Yokoe, D.; Kato, T.D. Mass transfer in Yb silicates under oxygen potential gradients at high temperatures. *AMTC Lett.* **2019**, *6*, 150–151.
13. Matsudaira, T.; Kitaoka, S.; Shibata, N.; Ikuhara, Y.; Takeuchi, M.; Ogawa, T. Effects of an oxygen potential gradient and water vapor on mass transfer in polycrystalline alumina at high temperatures. *Acta Mater.* **2018**, *151*, 21–30. [CrossRef]
14. Johnson, B.R.; Kriven, W.M.; Schneider, J. Crystal structure development during devitrification of quenched mullite. *J. Eur. Ceram. Soc.* **2001**, *21*, 2541–2562. [CrossRef]
15. Klemm, H. Silicon nitride for high-temperature applications. *J. Am. Ceram. Soc.* **2013**, *93*, 1501–1522. [CrossRef]
16. Suzuki, S.; Nasu, T.; Hayama, S.; Ozawa, M. Mechanical and thermal properties of β′-SiAlON prepared by slip casting method. *J. Am. Ceram. Soc.* **1996**, *79*, 1685–1688. [CrossRef]
17. Kitaoka, S.; Kawashima, N.; Suzuki, T.; Sugita, Y.; Shinohara, N.; Higuchi, T. Fabrication of continuous-SiC-fiber reinforced SiAlON-based ceramic composites by reactive melt infiltration. *J. Am. Ceram. Soc.* **2001**, *84*, 1945–1951. [CrossRef]

18. Wada, M.; Kitaoka, S.; Kawashima, N.; Yamada, T.; Yasutomi, Y.; Naito, K.; Koyama, M. Microstructure and thermal shock resistance of molten glass coated carbon materials fabricated by interfacial control. *J. Am. Ceram. Soc.* **2006**, *89*, 2134–2139. [CrossRef]

19. Kitaoka, S.; Matsushima, Y.; Chen, C.; Awaji, H. Thermal cyclic fatigue of porous ceramics for gas cleaning. *J. Am. Ceram. Soc.* **2004**, *87*, 906–913. [CrossRef]

20. Heuer, H.A.; Lou, L.K.V. Volatility diagrams for silica, silicon nitride, and silicon carbide and their application to high-temperature decomposition and oxidation. *J. Am. Ceram. Soc.* **1990**, *73*, 2785–3218. [CrossRef]

21. Jacobson, S.N. Corrosion of silicon-based ceramics in combustion environments. *J. Am. Ceram. Soc.* **1993**, *76*, 3–28. [CrossRef]

22. Schnurre, S.M.; Gröbner, J.; Schmid-Fetzer, R. Thermodynamics and phase stability in the Si–O system. *J. Non Cryst. Solids* **2004**, *336*, 1–25. [CrossRef]

Article

Characterization of Thermochemical and Thermomechanical Properties of Eyjafjallajökull Volcanic Ash Glass

Rebekah I. Webster [1], Narottam P. Bansal [2], Jonathan A. Salem [2], Elizabeth J. Opila [1] and Valerie L. Wiesner [3,*]

[1] Materials Science and Engineering Department, University of Virginia, Charlottesville, VA 22903, USA; riw5pv@virginia.edu (R.I.W.); ejo4n@virginia.edu (E.J.O.)

[2] Materials and Structures Division, NASA Glenn Research Center, Cleveland, OH 44135, USA; narottam.p.bansal@nasa.gov (N.P.B.); jonathan.a.salem@nasa.gov (J.A.S.)

[3] Advanced Materials and Processing Branch, NASA Langley Research Center, Hampton, VA 23681, USA

* Correspondence: valerie.l.wiesner@nasa.gov; Tel.: +1-757-864-4384; Fax: +1-757-864-8312

Received: 23 December 2019; Accepted: 19 January 2020; Published: 23 January 2020

Abstract: The properties of a volcanic ash glass obtained from the Eyjafjallajökull eruption of 2010 were studied. Crystallization experiments were carried out on bulk and powdered glass samples at temperatures between 900 and 1300 °C. Iron oxides, Fe_3O_4 and Fe_2O_3, and a silicate plagioclase, $(Na,Ca)(Si,Al)_4O_8$, were observed. Bulk samples remained mostly amorphous after up to 40 h at temperature. Powdered glass samples showed increased crystallinity after heat treatment compared to bulk samples. The average coefficient of thermal expansion of the glass was 7.00×10^{-6} K^{-1} over 25–720 °C. The Vickers hardness of the glass was 6–7 GPa and the indentation fracture toughness, 1–2 MPa $\sqrt{m}$ Values for density, elastic modulus, and Poisson's ratio were 2.52 g/cm^3, 75 GPa, and 0.24, respectively. The viscosity of the glass was determined experimentally and compared to three common models from the literature. The implications for the deposition of volcanic ash on hot section components of aircraft turbine engines are discussed.

Keywords: CMAS; EBCs; TBCs; volcanic ash; crystallization; thermal expansion; hardness; indentation fracture toughness; viscosity

---

1. Introduction

Particulates comprised of mainly calcium, magnesium, aluminum, and silicon oxides (CaO–MgO–Al$_2$O$_3$–SiO$_2$ or CMAS) stand as a challenge in the development of coating systems for next-generation aircraft turbine engines. CMAS originates as siliceous debris, such as sand, volcanic ash, or runway dust, which can be ingested into the turbine on takeoff, landing, or during flight and deposit on hot-section components [1,2]. CMAS melts at ~1200 °C and can infiltrate the protective barrier coatings (thermal and environmental barrier coatings, T/EBCs) needed for engine components.

Currently, superalloy components can reach a maximum temperature of ~1200–1300 °C due to the presence of a thermal barrier coating (TBC) which acts as a thermally protective topcoat [3]. Silicon carbide (SiC)-based ceramic matrix composites (CMCs) are proposed to replace some conventional Ni-base superalloys as turbine engine materials due to their higher operating temperatures and lighter weight, resulting in increased efficiency [4]. EBCs are required as a topcoat for CMCs to prevent thermally grown silicon oxide (SiO$_2$) from volatilizing in water vapor, a species resulting from the combustion of jet fuel in the engine. In addition to stability in water vapor and high-temperature environments, among other requirements [5], EBCs must be resistant to CMAS attack. The target operating temperature for CMCs approaches 1500 °C [6]. The temperatures experienced by both T/EBC systems are typically above CMAS melting temperatures [7].

It has been proposed in the literature that inducing crystallization of CMAS at the T/EBC-melt interface is a viable strategy in mitigating CMAS attack by limiting the extent of glass ingress [8,9]. Many studies have been conducted on different coating materials wherein the ability of the T/EBC to promote crystallization of CMAS was probed and the evolution of coating/glass phases was monitored as a function of temperature and/or time [8,10–13]. Conversely, few reports have isolated the crystallization behavior of CMAS alone. CMAS composition can vary widely, depending on its source and location [14]. As such, the intrinsic properties of these glasses may also be expected to differ. In addition to crystallization behavior, CMAS composition influences its melting temperature and viscosity, both important parameters when considering coating infiltration. Mechanical properties (i.e., coefficient of thermal expansion, hardness, toughness) of CMAS can also vary and are required to fully describe coating degradation.

Zaleski et al. used differential scanning calorimetry (DSC) to probe the melting and crystallization behavior of ternary CAS, CMAS, and Fe-containing CFAS and CMFAS synthetic deposits [15]. In general, it appeared that most of the melts under study were resistant to crystallization upon cooling at 10 °C/min. This behavior was most prominent in CAS compositions. The addition of Fe, however, promoted the crystallization of phases including hematite (Fe_2O_3), esseneite ($CaFeAlSiO_6$), and Ca-ferrite ($CaFe_2O_4$), as indicated by x-ray diffraction (XRD). Isothermal DSC measurements on a $Ca_{13}Fe_{10}Al_{18}Si_{59}$ (mol %) glass showed significantly increased crystallization kinetics compared to the ternary $Ca_{15}Al_{20}Si_{65}$ having a similar Ca:Si ratio and Al content. The presence of Fe is likely important when considering volcanic ash glasses, as bulk compositions have been previously reported with up to 16 at % Fe [16].

The objective of the current work was to characterize the intrinsic properties of a volcanic ash glass containing CMAS and Fe_2O_3, TiO_2, Na_2O, and K_2O. The crystallization behavior of the volcanic ash glass was determined in bulk and powdered form. Thermal and mechanical properties including melting (T_m) and glass transition (T_g) temperature, coefficient of thermal expansion (CTE), hardness, indentation fracture toughness, elastic modulus (E), and Poisson's ratio (ν) were also measured. Glass viscosity was determined experimentally and compared to calculated values based on composition using several viscosity models described in the literature. The intrinsic properties of the volcanic ash glass were compared to those of previously investigated synthetic sand and desert sand CMAS glasses [17–19]. The significance of results with respect to T/EBC design are discussed.

2. Experimental Procedure

Ash obtained from the 2010 eruption of Iceland's Eyjafjallajökull volcano was melted at 1500 °C for 1 h in a platinum crucible followed by quenching in water. The composition of the resulting glass was determined by inductively coupled plasma atomic emission spectroscopy (ICP-AES) performed by a commercial test laboratory and can be found in Table 1. Table 1 also includes the composition of a synthetic sand glass [17,18] and a desert sand glass [19] for comparison.

Differential thermal analysis (DTA) was carried out using a Netzsch STA 409 (Burlington, MA, USA) on bulk and powder glass samples to determine melting and crystallization behavior. Scans were performed in flowing air from room temperature to 1500 °C at a ramp rate of 10 °C/min. A platinum pan was used to contain each sample. Prior to each run, a background scan was performed under identical conditions and was subsequently subtracted from the sample scan.

Isothermal heat treatments were carried out in a stagnant air box furnace on bulk glass pieces weighing approximately 250–350 mg. Samples were placed on a platinum foil and were either exposed for 1, 10, or 20 h at 900 and 1000 °C (low-temperature furnace; Neytech Vulcan 3-550; Bloomfield, NJ, USA) or for 1, 10, 20, or 40 h at 1100, 1150, 1200, and 1300 °C (high-temperature furnace; Carbolite HTF 18/27; Hope Valley, UK). Samples were cooled at either 10 °C/min or quenched in air. Some experiments were carried out using glass powder instead of bulk specimens; in these cases, bulk pieces weighing 250–350 mg were ground with mortar and pestle to a fine grit prior to exposure. After heat treatment, samples were either crushed for x-ray diffraction (XRD) phase analysis (Bruker D8 Advance; Billerica, MA, USA) or mounted and polished in cross-section for imaging by scanning electron microscopy

(SEM—Phenom-World Phenom Pro; Eindhoven, The Netherlands). Mounted and polished specimens were coated with a thin layer of platinum prior to imaging. Quantitative XRD was performed by mixing 20–25 wt % α-Al_2O_3 (corundum) with the powdered samples after heat treatment. Powders were mixed thoroughly by grinding with a mortar and pestle. Quantitative XRD scans were run from a 2θ of 10° to 120° with a step size of 0.02° and a step time of 2.5 s/step. Sample peaks were referenced to the Al_2O_3 standard in the Whole Pattern Fitting (WPF) module of Jade 2010 software from Materials Data Inc. (MDI; Livermore, CA, USA).

Table 1. Composition (in mol % and wt %, determined by inductively coupled plasma atomic emission spectroscopy—ICP-AES) of the volcanic ash glass prepared in this study compared to previously investigated synthetic sand and desert sand glasses [17–19]. Estimated uncertainty for values in the range of 1.0–3.0% = ±10% of actual value, for values in the range of 3.0–10.0% = ±5% of actual value, for values in the range of 10.0–25.0% = ±2% of actual value, for values in the range of 25.0–75.0% = ±1% of actual value.

		CaO	MgO	Al_2O_3	SiO_2	Fe_2O_3	TiO_2	Na_2O	K_2O	Trace Oxides
Volcanic ash glass	mol %	6.2	4.2	10.1	67.5	4.4	1.4	5.0	1.2	–
	wt %	4.9	2.4	14.6	57.3	10.0	1.6	4.4	1.6	bal.
Synthetic sand glass	mol %	23.3	6.4	3.1	62.5	0.04	–	4.1	0.5	–
	wt %	21.9	4.3	5.4	63	0.1	–	4.3	0.8	bal.
Desert sand glass	mol %	27.8	4	5	61.6	0.6	–	–	1	–
	wt %	25.2	2.6	8.2	59.8	1.6	–	–	1.5	bal.

Sample discs were prepared for mechanical testing and dilatometry using a Centorr mini hot press (Nashua, NH, USA). Glass powder was loaded into a graphite die and treated at 600 °C and 2 ksi (13.8 MPa) for 10 min under vacuum. Specimens were cooled to room temperature at a rate of 10 °C/min after the applied pressure was removed. Graphite foil was used to cover the die pieces in contact with the glass during pressing, and any residual foil was removed from sample surfaces by polishing.

Glass density was calculated from the measured mass and volume of the hot-pressed glass disc. The impulse excitation method described in ASTM C1259 [20] was used to determine the elastic modulus and Poisson's ratio of the glass at room temperature. An Audio Technica ATM350 condenser microphone with an M-Audio DMP preamplifier was used to acoustically measure and amplify the natural frequency of the glass disc upon mechanical excitation in the desired mode. The acoustic signal acquisition hardware and Sound & Vibration Toolset software from National Instruments (Austin, TX, USA) were used to determine the disc's frequency, which was used to calculate the elastic modulus and Poisson's ratio.

A Struer's DuraScan (Cleveland, OH, USA) was used to apply a Vickers diamond indent onto the glass disc surface, which was polished according to ASTM C1327-08 [21]. Three indentations were made on the polished disc at each individual load of 1.96, 2.94, 4.9, and 9.8 N applied for 15 s. The diagonal (2a) and crack (2c) lengths of each indentation were measured.

A glass bar with length of 2.5 cm was evaluated using a Netzsch differential dilatometer model 402 C (Burlington, MA, USA) interfaced with a computerized data acquisition and analysis system to measure the glass transition temperature (T_g), softening point (T_d), and linear coefficient of thermal expansion (CTE; α). The glass bar was heated at a rate of 5 °C/min in the air from room temperature to 1000 °C.

Viscosity measurements were performed using an Orton RSV-1600 viscometer furnace setup (Westerville, OH, USA) equipped with a Brookfield (Middleboro, MA, USA) HA-DV2T viscosity measuring unit and platinum spindle. A 50 mL platinum crucible was filled with approximately 80 g of volcanic ash glass for a molten glass volume of approximately 30 mL. Viscosity measurements were performed between about 1300 and 1500 °C. The viscometer furnace was held at 1500 °C for at least

20 min to equilibrate the melt. Two methods were utilized for collecting data. In the first, the spindle was lowered into the melt and allowed to rotate continuously as the furnace cooled at 2 °C/min. Viscosity data was collected during cooling. In the second, the spindle was lowered into the melt and allowed to rotate. The furnace was cooled at 50 °C intervals and held at each temperature for 20 min. The viscosity of the glass was determined at each temperature after a stable reading was obtained.

The viscosity of the glass melt was determined by measuring the percent spindle torque required to maintain the spindle rotation at the desired speed. The viscosity of the melt (η; in centipoise) is related to the spindle torque (τ) and rotational speed (RPM) by the following equation, found in the Orton RSV-1600 instrument manual:

$$\eta = \frac{100}{RPM} \times TK \times SMC \times \tau \tag{1}$$

where TK and SMC are the viscometer torque constant and the spindle multiplier constant, respectively. These values were provided in the viscometer software. The viscosity of a borosilicate glass standard (NIST SRM 717a, Sigma Aldrich; St. Louis, MO, USA) was determined experimentally between 1000–1325 °C, confirming that the supplied constants gave accurate data in the temperature region of interest (Figure 1). The borosilicate glass standard was not run above 1325 °C due to volatility concerns. The calculated viscosity curve (Figure 1) was determined using viscosity constants provided by NIST.

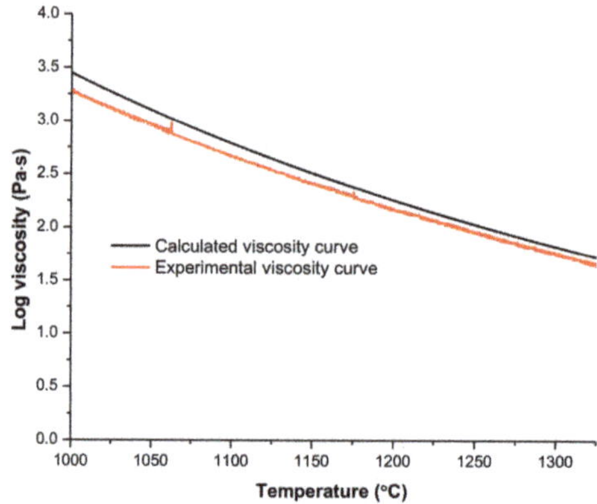

Figure 1. Calculated and experimental viscosity curves for a borosilicate glass standard.

3. Results

3.1. Crystallization Behavior—DTA

DTA heating and cooling curves at a ramp rate of 10 °C/min are given in Figure 2 for the bulk and powdered glass. A list of observed DTA thermal events, described for bulk and powdered glasses, is given in Table 2. On heating, the bulk glass exhibited exothermic peaks at approximately 900 (Figure 2—1$_B$) and 1100 °C (2$_B$), which corresponded to crystallization. Each peak is likely attributed to the crystallization of a distinct phase. There appeared to be a melt endotherm with a peak temperature of approximately 1350 °C; however, the peak itself was very broad. A dip in the baseline was evident at about 1050 °C (3$_B$), suggesting that this glass started to melt around 1050 °C and might not have a distinct melting temperature. Indeed, in separate box furnace experiments glass specimens were observed to bead by 1100 °C and wet the platinum holder substrate by 1150 °C. Bulk glass specimens after heat treatment at 1100 and 1150 °C are shown in Figure 3. Bulk glass heat treated at 1300 °C for 1 h and air quenched was almost entirely amorphous by XRD analysis, suggesting that by this temperature, the glass was nearly completely molten. On cooling, an exothermic peak was also observed around

900 °C (4_B) suggesting that a similar phase that crystallized around 900 °C on heating (1_B) might have also evolved upon cooling at 10 °C/min.

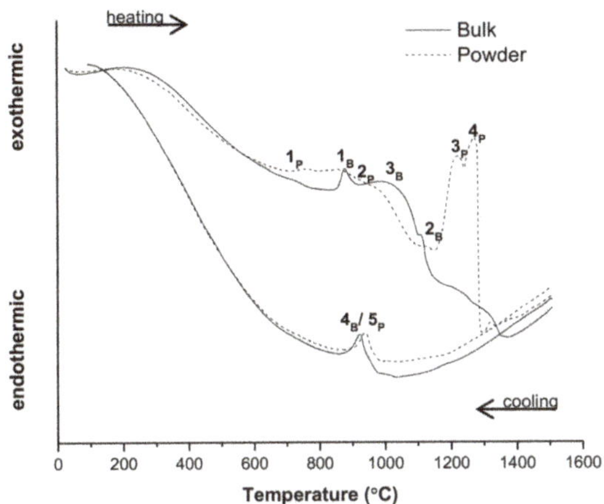

Figure 2. Bulk (bold line) and powder (dashed line) volcanic ash glass differential thermal analysis (DTA) curves with a ramp rate of 10 °C/min. Labels 1_B, 2_B, 1_P, 2_P, and 3_P correspond to crystallization in the bulk (B) and powder (P) samples on heating. Labels 3_B and 4_P show the onset of melting. Labels 4_B and 5_P indicate solidification of molten glass upon cooling.

Table 2. List of DTA thermal events presented in Figure 2 for the volcanic ash glass.

	Bulk	Approximate Temperature (°C)	Powder	Approximate Temperature (°C)
On heating				
Onset Melting (endo)	3_B	1050 °C (peak melting at 1350 °C)	4_P	950–1000 °C
Crystallization (exo)	1_B, 2_B	900, 1100 °C	1_P, 2_P, 3_P	900, 1225, 1275 °C
On cooling				
Crystallization (exo)	4_B	900 °C	5_P	900 °C

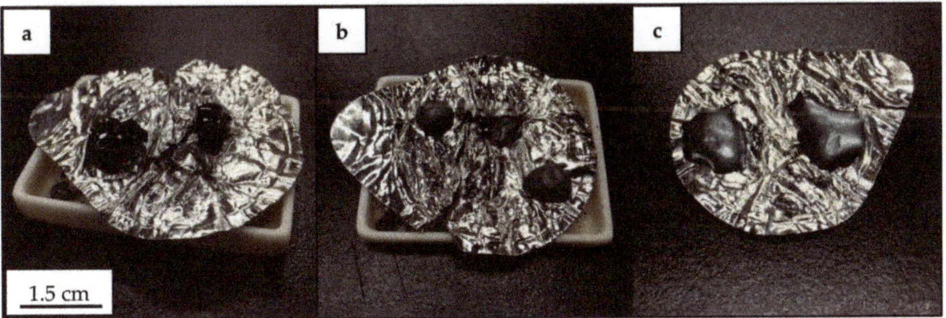

Figure 3. Bulk volcanic ash glass samples before heat treatment (**a**) and after heat treatment at 1100 (**b**) and 1150 °C (**c**). The solidified glass samples are the dark droplets on Pt foil.

The heating curve for powdered glass was quite different from that seen for the bulk, as shown in Figure 2. There was a broad exothermic hump that centered around 900 °C. This peak was not as distinct as for the bulk glass and had a comparatively lower onset temperature (1_P). In addition, instead of having a second exothermic peak around 1100 °C, there were two very large peaks at approximately

1225 (2_P) and 1275 °C (3_P). It is somewhat difficult to distinguish background information from actual thermal events for this curve, however, the melting behavior appeared similar to that for the bulk—the onset (4_P) and peak melting temperatures being shifted to slightly lower values. On cooling, there was an exothermic peak at approximately 900 °C (5_P), echoing that observed for the bulk sample.

3.2. Crystallization Behavior—Box Furnace Experiments

XRD plots for bulk glass pieces treated for 1 h at 900–1200 °C can be seen in Figure 4. All scans were collected for samples after being heated and cooled at 10 °C/min. XRD patterns for specimens that were placed in and removed from the furnace at temperature (i.e., quenched in air) did not show differences in phase or amount of phase present when compared to samples that were ramped to and from temperature. Consequently, XRD results for quenched specimens were excluded from the rest of this report. In Figure 4, the XRD scan for glass not exposed to temperature is also shown (labeled "Control") for comparison. Only two phases were discerned—Fe_3O_4 (magnetite) and Fe_2O_3 (hematite). After heat treatment at 900 and 1000 °C, broad, shallow peaks for magnetite were present, but the glass appeared to have remained mostly amorphous. At 1100, 1150, and 1200 °C, the peak for magnetite appeared larger and sharper. At 1150 and 1200 °C, the 100% intensity peak for hematite began to emerge.

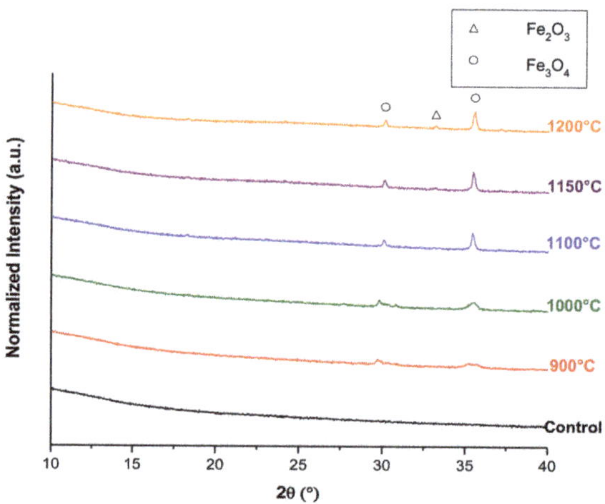

Figure 4. XRD plots for bulk volcanic ash glass heat treated for 1 h at 900, 1000, 1100, 1150, and 1200 °C (ramp rate 10 °C/min). The "Control" spectrum is for volcanic ash glass not exposed to temperature.

As the amount of time held at temperature was increased for 1100–1200 °C treatments, the magnetite was converted to hematite (Table 3). The transition from magnetite to hematite occurred more rapidly as the temperature increased, such that there was a higher relative amount of hematite after 10 h at 1200 °C compared to after 10 h at 1100 °C (Table 3). This is likely due to the enhanced diffusion of oxygen through the bulk glass at higher temperatures [22]. The reaction of Fe_3O_4 to Fe_2O_3 is given by [23]

$$4Fe_3O_4 + O_2(g) \leftrightarrow 6Fe_2O_3 \tag{2}$$

The evolution of the microstructure of samples held at 1150 °C can be seen in Figure 5. Energy dispersive spectroscopy (EDS) was unable to differentiate Fe_3O_4 from Fe_2O_3. The bright (by backscattered electron imaging, BSE) Fe-based precipitates appear as both faceted particulates and rod-like structures. Faceted particles are mostly submicron in size after 1 h at temperature but grow to a maximum diameter of approximately 10 μm with increasing time. In some areas, the surface of the glass had a thin layer or "shell" of Fe-based particulates (somewhat visible in Figure 5c,d, more apparent in Figure 6).

Table 3. The phase composition of bulk volcanic ash glass samples after heat treatment using quantitative XRD analysis.

	wt % Amorphous	wt % Fe_3O_4	wt % Fe_2O_3	wt % Plagioclase
1100 °C				
1 h	88.7 ± 0.1	9.9 ± 0.1	1.4 ± 0.1	0.0
10 h	87.8 ± 0.2	8.9 ± 0.1	1.0 ± 0.1	2.3 ± 0.2
20 h	89.4 ± 0.1	5.9 ± 0.1	2.3 ± 0.1	2.4 ± 0.1
40 h	81.7 ± 0.3	4.6 ± 0.1	3.0 ± 0.1	10.7 ± 0.3
1150 °C				
1 h	90.7 ± 0.1	8.5 ± 0.1	0.8 ± 0.0	0.0
10 h	94.2 ± 0.1	2.8 ± 0.0	3.0 ± 0.0	0.0
20 h	94.5 ± 0.1	1.6 ± 0.1	3.9 ± 0.0	0.0
40 h	95.7 ± 0.0	0.0	4.3 ± 0.0	0.0
1200 °C				
1 h	90.6 ± 0.1	7.7 ± 0.1	1.8 ± 0.1	0.0
10 h	95.5 ± 0.1	0.9 ± 0.1	3.6 ± 0.0	0.0
20 h	95.4 ± 0.1	1.4 ± 0.0	3.2 ± 0.0	0.0

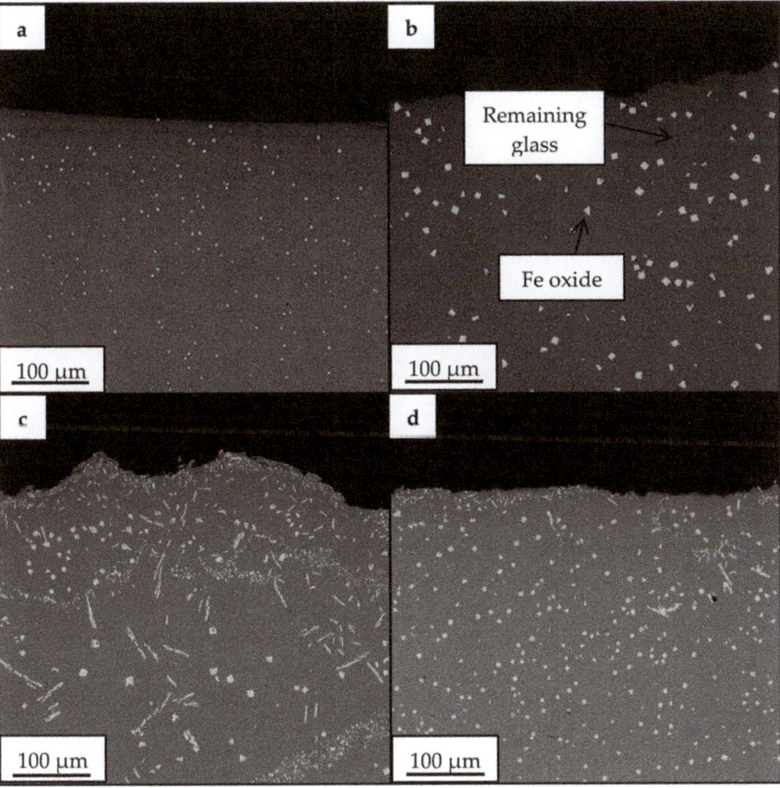

Figure 5. Cross-section backscattered electron imaging (BSE) images of bulk volcanic ash glass after exposure at 1150 °C for 1 h (**a**), 10 h (**b**), 20 h (**c**), and 40 h (**d**).

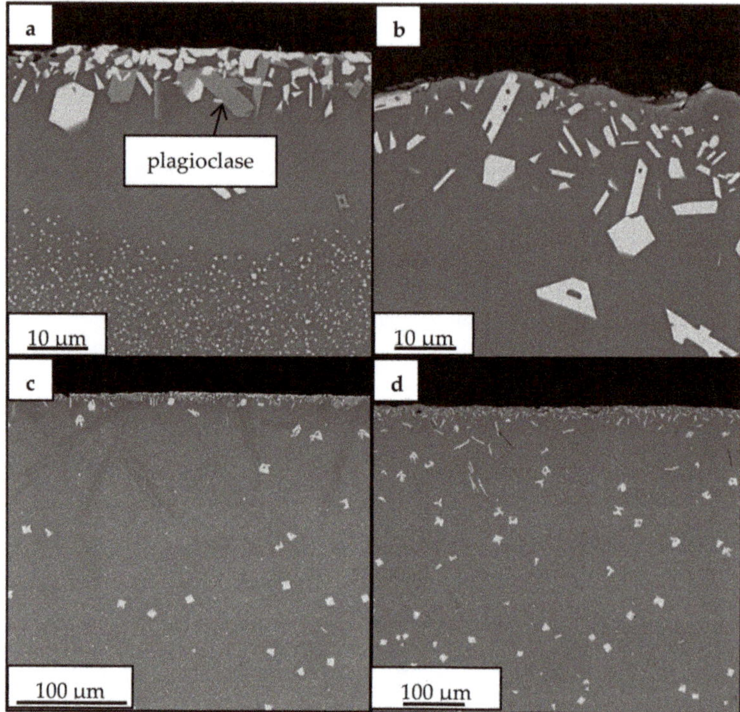

Figure 6. Cross-section BSE images for bulk volcanic ash glass samples treated at 1100 (**a,c**) and 1150 °C (**b,d**) for 10 h.

At 1100 °C, a third phase, likely a solid solution of anorthite ($CaAl_2Si_2O_8$) and albite ($NaAlSi_3O_8$), was detected after holding for 10 h (see Figure 6). This phase (hereto referred to as plagioclase) may also be present in small amounts at 1000 °C.

SEM BSE cross-section images are presented in Figure 6 for samples held at 1100 and 1150 °C for 10 h. The results given by SEM supported those of XRD. After 10 h at 1100 °C, there appeared to be two phases growing at the surface of the glass (Figure 6a). EDS confirmed a Fe-rich phase (lighter contrast by BSE) and a Si-rich phase (darker by BSE). Iron oxides (likely mostly Fe_3O_4 based on XRD) were also present throughout the bulk of the sample (Figure 6c) but were generally smaller than those in the 1150 °C sample (Figures 5b and 6d). A thin layer of concentrated Fe precipitates can be seen at the surface of both the 1100 and 1150 °C (Figure 6c,d) samples. This "shell" is likely the Fe_2O_3 phase, as it has been previously observed in magnetite iron ore pellets oxidized at 800–1200 °C [24]. The "shell" was not always continuous across the surface of samples, which may be due to the removal of material when separating the bulk glass from its platinum substrate. As expected from XRD, no plagioclase phase was observed in the 1150 °C sample (Figure 6b).

When powdered samples were heated in the box furnace at 1100 and 1150 °C, the prominent crystallized phases were Fe_2O_3 and plagioclase (Figure 7). The amounts of each phase are given in Table 4. It is expected that the increased amount of crystallized glass compared to bulk samples is due to the increased surface area of the powder.

FactSage free energy minimization calculations were performed for both the volcanic ash glass and the synthetic sand glass compositions (Table 1), using the Equilibrium Module and FToxide database, including SLAGA [25], to determine the expected phase distributions at a given temperature based on thermodynamics alone. Respective amounts of the different CMAS oxides (in mol %, Table 1) were used as FactSage inputs. Calculations were performed at 1 atm. As was observed experimentally,

FactSage indicated that Fe_2O_3 and albite ($NaAlSi_3O_8$) and anorthite ($CaAl_2Si_2O_8$) were among the expected phase formations. Between 900–1150 °C, SiO_2, $MgSiO_3$, $KAlSi_3O_8$, $CaSiTiO_5$, and $CaMgSi_2O_6$ (diopside) were among other phases also predicted to appear. Albite was not seen above 1050 °C and Fe_2O_3 was the only expected solid-phase above ~1200 °C. The amount of the observed phases as a function of temperature is given in Figure 8. A Scheil–Gulliver cooling profile was also performed, with a start and stop temperature of 1300 and 900 °C, respectively, and a step size of 25 °C. This method subtracts precipitated phase constituents from the overall glass composition as they form. At and above 1200 °C, only Fe_2O_3 is expected. Below 1200 °C, anorthite forms; at ≤1100 °C, $MgSiO_3$, diopside, and $CaSiTiO_5$ appear in addition to Fe_2O_3 and anorthite. Albite is only discerned at ≤1000 °C.

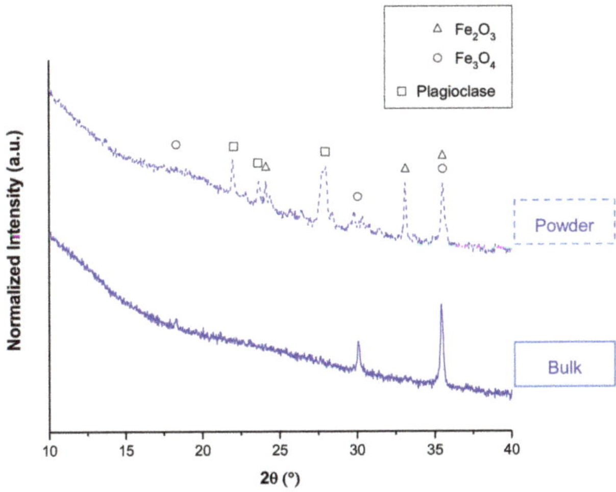

Figure 7. XRD spectra for bulk (solid line) and powder (dashed line) volcanic ash glass samples exposed to 1100 °C for 1 h.

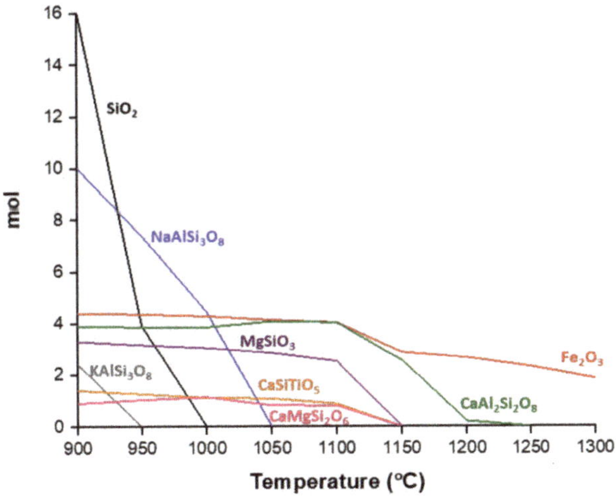

Figure 8. Expected phase formations in the volcanic ash glass as a function of temperature, calculated using the FactSage Equilibrium module.

The same processes were performed for the synthetic sand glass. Wollastonite (CaSiO$_3$) and diopside were among the expected phase formations between 900–1300 °C. Other predicted phases (between 900–1000 °C) included albite, SiO$_2$, and Na$_2$Ca$_3$Si$_6$O$_{16}$. Figure 9 plots the expected phase formations in the synthetic sand glass as a function of temperature.

Table 4. The phase composition of powder volcanic ash glass samples after heat treatment using quantitative XRD analysis.

	wt % Amorphous	wt % Fe$_3$O$_4$	wt % Fe$_2$O$_3$	wt % Plagioclase
1100 °C				
1 h	50.6 ± 0.5	3.9 ± 0.1	7.2 ± 0.1	38.3 ± 0.5
1150 °C				
1 h	65.5 ± 0.3	0.5 ± 0.1	7.1 ± 0.1	26.8 ± 0.3

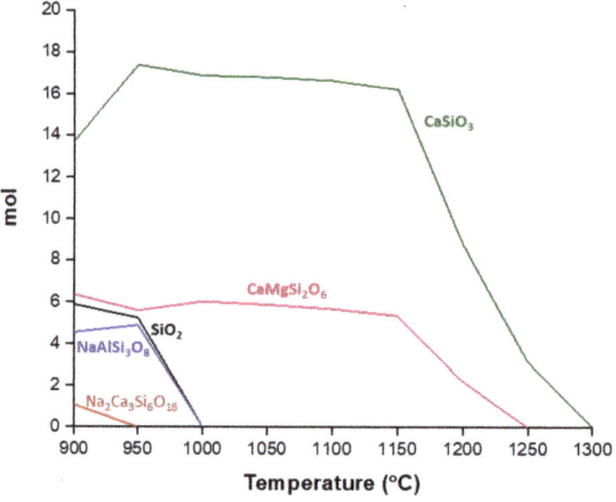

Figure 9. Expected phase formations in the synthetic sand glass as a function of temperature, calculated using the FactSage Equilibrium module.

3.3. Thermal Behavior—Dilatometry

Figure 10 shows the dilatometric thermal expansion curve for the hot-pressed volcanic ash glass bar heated at 5 °C/min from RT to 1000 °C. Dilatometric values for the volcanic ash glass, synthetic sand glass, and desert sand glass are found in Table 5. The T_g and T_d of the volcanic ash glass were observed to be 741 and 844 °C, respectively. Comparatively, the synthetic and desert sand glasses referred to in Table 1 had reported T_g values of 694 and 706 °C, respectively, and T_d values of 751 and 764 °C, respectively. The average linear CTE for the bulk volcanic ash glass was 7.00 × 10^{-6} K^{-1} between 25–720 °C. This value was lower than the reported ~9–10 × 10^{-6} K^{-1} for the synthetic and desert sand glasses over the same temperature range.

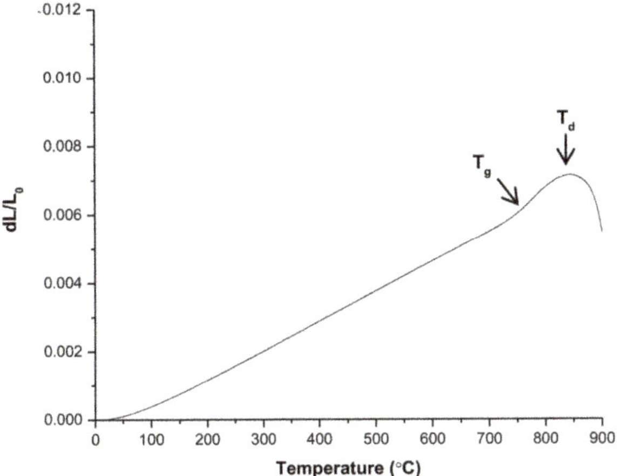

Figure 10. Dilatometric thermal expansion curve on heating for the volcanic ash glass at a heating rate of 5 °C/min in air. Glass transition temperature (T_g) and glass softening temperature (T_d) are indicated.

Table 5. Thermal and mechanical properties of volcanic ash, synthetic sand, and desert sand glasses.

Property	Volcanic Ash Glass	Synthetic Sand Glass *	Desert Sand Glass ±
Melting (T_m, °C)	~1300–1350 °C	1176	–
Density (g/cm^3)	2.52	2.63	2.69
Glass transition temperature (T_g, °C)	741	694	706
Dilatometric softening point (T_d, °C)	844	751	764
Coefficient of thermal expansion (CTE α, K^{-1})	7.00×10^{-6} (~25–720 °C)	9.32×10^{-6} (~25–690 °C)	9.8×10^{-6} (~25–700 °C)
Young's Modulus (E, GPa)	75	84.3	92.3
Poisson's ratio (ν)	0.24	0.26	0.28

* Values obtained from [17,18]; ± Values obtained from [19].

3.4. Density and Mechanical Properties

The bulk density, Elastic Modulus (E), Poisson's ratio (ν), Vickers hardness (H_V), and indentation fracture toughness (evaluated by three different equations) for the volcanic ash glass are reported in Tables 5 and 6. Density was found to be 2.52 g cm^{-3} and E and ν to be 75 GPa and 0.24, respectively. These values were comparable to those reported for the synthetic and desert sand glasses, though E for the synthetic and desert sand glasses was approximately 84 and 92 GPa, respectively, indicating that the volcanic ash glass was less stiff.

Vickers microhardness was determined at each load using the following equation [21]:

$$H_V = 1.8544 \left[\frac{P}{(2a)^2} \right] \tag{3}$$

where P was the applied load and $2a$ was the indentation diagonal length. Vickers hardness values determined at loads of 1.96, 2.94, 4.9, and 9.8 N are reported in Table 6.

Indentation fracture toughness (K_C) of the volcanic ash glass was calculated from indentation load and crack and diagonal lengths using relationships found in the literature. Miyoshi et al. [26] used the following:

$$K_C = 0.026aE^{0.5} P^{0.5} c^{-1.5} \tag{4}$$

where E was the Young's Modulus, P was the indentation load, and a and c were half the indent length and crack length, respectively. Marshall and Evans [27] evaluated indentation fracture toughness using the following relation:

$$K_C = 0.036E^{0.4} P^{0.6} a^{0.8} c^{-1.5} \tag{5}$$

Anstis et al. [28] presented:

$$K_C = (0.016 \pm 0.004)\, P\left(\frac{E}{H_v}\right)^{0.5} c^{-1.5} \tag{6}$$

to determine the indentation fracture toughness of glass specimens. K_C values calculated using Equations (4)–(6) can be found in Table 6 for a range of indentation loads. The loads were chosen to prevent spalling while giving a reasonably long crack ($c > 2a$). The hardness values for the volcanic ash glass are comparable to those of the synthetic and desert sand glasses, but the volcanic ash glass is significantly tougher (~1–2 MPa $\sqrt{m}$). The toughness of the synthetic and desert sand glasses is akin to that of many glasses, however, that of the volcanic ash glass more closely matches a glass-ceramic [29]. Hot pressing the glass at 600 °C resulted in the crystallization of some Fe_3O_4 precipitates in the sample (data not shown), which may explain the increased toughness.

Table 6. Vickers hardness and indentation fracture toughness at various indent loads for the volcanic ash glass and synthetic sand and desert sand glasses.

Indent Load, P (N)	Vickers Hardness, H_V (GPa)	Indentation Fracture Toughness, K_C (MPa $\sqrt{m}$) Calculated Using Equations (3)–(5)		
		Miyoshi et al. [26]	Marshall and Evans [27]	Anstis et al. [28]
Volcanic ash glass (current study)				
1.96	7.02 ± 0.48	n/a	n/a	n/a
2.94	6.22 ± 0.28	1.60 ± 0.24	1.83 ± 0.28	1.43 ± 0.22
4.9	6.99 ± 0.35	1.31 ± 0.08	1.53 ± 0.09	1.17 ± 0.07
9.8	7.03 ± 0.37	1.26 ± 0.13	1.47 ± 0.15	1.12 ± 0.12
Grand mean	6.75 ± 0.40	1.39 ± 0.18	1.61 ± 0.19	1.24 ± 0.17
Synthetic sand glass [17,18]				
1.96	6.12 ± 0.19	0.72 ± 0.04	0.82 ± 0.04	0.62 ± 0.04
2.94	6.28 ± 0.19	0.74 ± 0.05	0.86 ± 0.05	0.68 ± 0.04
4.9	6.04 ± 0.18	0.66 ± 0.05	0.74 ± 0.05	0.6
9.8	6.12 ± 0.13	0.62 ± 0.04	0.72 ± 0.04	0.58 ± 0.04
Grand mean	6.14 ± 0.10	0.69 ± 0.06	0.79 ± 0.07	0.62 ± 0.04
Desert sand glass [19]				
1.96	5.90 ± 0.10	0.70 ± 0.03	0.70 ± 0.03	0.60 ± 0.03
2.94	6.60 ± 0.10	0.80 ± 0.05	0.90 ± 0.10	0.70 ± 0.05
4.9	6.40 ± 0.10	0.75 ± 0.05	0.80 ± 0.10	0.65 ± 0.05
9.8	6.20 ± 0.10	0.65 ± 0.05	0.70 ± 0.10	0.60 ± 0.10
Grand mean	6.28 ± 0.30	0.73 ± 0.06	0.78 ± 0.10	0.64 ± 0.05

3.5. Viscosity

Experimental viscosity values for the volcanic ash glass, obtained by both continuous cooling and isothermal holds, are plotted in Figure 11 as a function of temperature. EDS analysis of the glass following viscosity testing suggested that the composition of the glass did not change during testing.

The experimental viscosities range from ~15–105 Pa·s (log values ~1.25–2) with a decreasing temperature. These values are considerably lower than expected from model predictions. Glass viscosity was predicted using three different viscosity models (Giordano et al., Fluegel, and the FactSage Viscosity Module with Melt Database). The Giordano et al. model [30] calculates the non-Arrhenian temperature dependence of viscosity for naturally occurring silicate melts by connecting experimentally obtained viscosity profiles with VFT (Vogel–Fulcher–Tamman) constants. The VFT equation is given by [31–33]

$$\log \eta = A + \frac{B}{T - C} \tag{7}$$

where A, B, and C are constants. The Fluegel model [34] also correlates VFT constants to experimental viscosity data, but for commercial silicate-based glasses. On the other hand, the FactSage method [25] of calculation is non-empirical, instead relating viscosity to melt structure. The Modified Quasichemical Model is utilized, along with thermodynamic values, to calculate viscosity for a given composition. The viscosity, based off of the Giordano et al., Fluegel, and FactSage models, is given as a function of temperature for the volcanic ash glass in Figure 11, alongside experimental values. It is assumed that the glass is fully molten at temperatures 1325–1500 °C based on the results discussed previously.

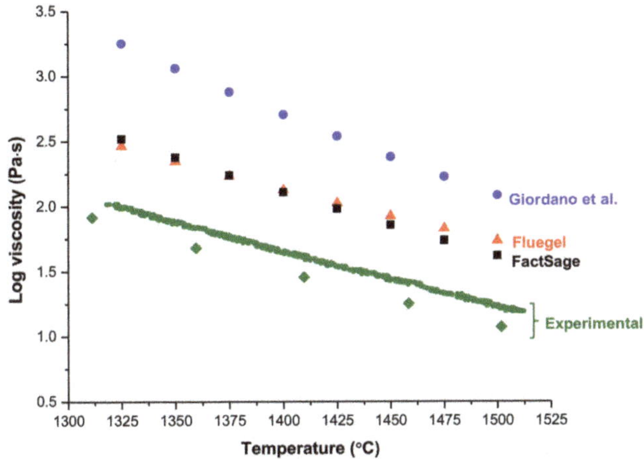

Figure 11. Viscosity curves for the volcanic ash glass used in this study based on experimental and calculated values. Calculated values were determined using the FactSage Viscosity Module and Melt Database (black squares), the Giordano et al. model (blue circles), and the Fluegel model (red triangles). Experimental data are given in green. The continuous curve is for the experimental method in which the glass was cooled at a constant 2 °C/min while the separate diamond-shaped data points are for the experimental method in which the glass was equilibrated at 50 °C intervals.

It can be seen from Figure 11 that the FactSage and Fluegel models give values that are in good agreement, while the Giordano model viscosity is greater by up to about one order of magnitude. Wiesner et al. performed similar calculations on the synthetic sand glass composition given in Table 1 and also saw that the FactSage and Fluegel models showed reasonable agreement. Experimental measurements confirmed that the FactSage and Fluegel models were more accurate than the Giordano et al. model for that particular composition [35,36]. This result did not translate to the volcanic ash glass, which had an experimental viscosity lower than the FactSage and Fluegel models by about half an order of magnitude.

It is important to note that the input composition for the Giordano model is slightly different from that reported in Table 1 due to model constraints—notably, that the input was FeO instead of Fe_2O_3. The composition of the glass for this model was $5.1CaO$-$2.5MgO$-$15.2Al_2O_3$-$59.9SiO_2$-$9.3FeO$-$1.7TiO_2$-$4.6Na_2O$-$1.7K_2O$ (wt %). The Fe_2O_3 reported in Table 1 was converted to FeO (assuming 2 mol FeO for 1 mol Fe_2O_3 based on Fe alone) and trace oxides were not taken into account due to their total weight being unknown.

Wiesner and Bansal [18] estimated the amount of time (t) it would take for their synthetic sand CMAS to infiltrate a 200 μm TBC, barring CMAS crystallization and/or other thermochemical interactions with the coating, using the following

$$t \sim \left[\frac{k_t}{8D_c} \left(\frac{1-\omega}{\omega} \right)^2 L^2 \right] \frac{\eta}{\sigma_{LV}} \qquad (8)$$

where a tortuosity (k_t) value of ~3 was considered for the coating [5], the capillary diameter (D_c) was 1 μm [37], $\omega \approx 0.1$ was the pore fraction open to flow [37], and surface tension (σ_{LV}) was assumed to be

~0.4 J m^{-2}, calculated using an approach for silicate glass melts at 1400 °C [38]. Assuming these values for the current system and using the viscosities calculated at 1400 °C by each model, the infiltration time was determined to be ~29 min for the Giordano model and ~7–8 min for both the FactSage and Fluegel models. These values are compared to the calculated infiltration times for the synthetic sand glass in Table 7. Given that the experimental viscosity for the volcanic ash glass is about half an order of magnitude less than that calculated by the FactSage and Fluegel approaches, the expected time to complete infiltration at 1400 °C is actually around 2 min. Experimental viscosities and infiltration times at ~1400 °C are also given in Table 7 for the volcanic ash glass and the synthetic sand glass reported previously [35].

Table 7. Viscosity (η) values (in Pa·s) as a function of temperature, model, and glass composition. The predicted time to complete the infiltration of a 200 µm thick thermal barrier coating (TBC) (dependent on viscosity) at 1400 °C is given for synthetic sand [35,36] and volcanic ash glasses. The expected infiltration time based on experimental viscosity data (using isothermal holds) is also given.

	Synthetic Sand				Volcanic Ash			
	Giordano	Fluegel	FactSage	Experimental	Giordano	Fluegel	FactSage	Experimental
η 1300 °C	274.9	33.6	78.1	19.5 (1318 °C)	2848	389	465	82.3 (1311 °C)
η 1400 °C	70.3	11.5	19.6	7.1 (1411 °C)	507.3	134.0	127.9	28.6 (1410 °C)
η 1500 °C	22.3	4.8	6.2	3.3 (1519 °C)	121.4	54.4	41.7	11.8 (1502 °C)
Infiltration Time (1400 °C)	3.8 min	*n/a*	1.1 min	0.41 min (1411 °C)	29.2 min	7.7 min	7.3 min	1.6 min (1410 °C)

4. Discussion

The crystallized phases observed in this study are similar to those reported elsewhere. Mechnich et al. previously investigated a synthetic volcanic ash glass, prepared by sol-gel, with a composition approximating that of an actual Eyjafjallajökull deposit [39]. Calcination of the base glass powder at 800 °C resulted in the formation of Fe_2O_3, as well as some pseudobrookite (Fe_2TiO_5) and aegirine ($NaFeSi_2O_6$), within the glass. After heat treatment of a powdered mixture of the volcanic ash glass with TBC yttria-stabilized zirconia (YSZ) at temperatures of 900–1200 °C, the hematite phase was still discerned and plagioclase was evident ≥1000 °C. An electron beam-physical vapor deposited (EB-PVD) YSZ coating exposed to the glass for 1 h at 1200 °C also showed the formation of plagioclase and an Fe-rich oxide at its surface. Jang et al. characterized an as-received volcanic ash deposit from the Eyjafjallajökull 2010 eruption and determined the presence of anorthite/albite (plagioclase), augite ($Ca(Mg,Fe,Al)(Si,Al)_2O_6$), and analcime ($Na(Al,Si_2O_6)H_2O$) [40]. Interestingly, they did not report any Fe-based phases in the as-received sample despite it being composed of nearly 10 wt % Fe_2O_3. The heat treatment history of their sample is unknown.

The volcanic ash glass showed considerably different crystallization behavior compared to a previously investigated synthetic sand glass (Table 1). Wiesner and Bansal [17] saw the formation of wollastonite ($CaSiO_3$) and aluminum diopside ($Ca(Mg,Al)(Si,Al)_2O_6$) in bulk CMAS samples at temperatures as low as 925 °C. For the glass in this study, while DTA indicated crystallization at 900 °C, and FactSage calculations suggested the same, furnace tests revealed that only a very small amount of Fe_3O_4 formed at this temperature. The synthetic sand CMAS glass was fully crystalline after 20 h at 925 °C or 5 h at 960 °C. Based on the results of this study, it is clear that bulk volcanic ash glass pieces would not fully crystallize at these temperature/time scales. It is interesting to note that the powderized glass showed significantly increased crystallization kinetics (Table 4) compared to the bulk (Table 3) at 1100 and 1150 °C. It is likely that the glass powder would also show increased crystallization at 900, 1000, and 1200 °C compared to the bulk. It is possible, and apparently more desirable in terms of crystallization, that a protective coating (T/EBC) in service will come into contact with fine particles instead of bulk pieces.

Volcanic ash glass of this composition in contact with a T/EBC at temperatures nearing 1300 °C will likely soften/melt and penetrate the coating via defects such as open channels (in a TBC) or grain boundaries and pores/cracks (in an air plasma spray (APS)-deposited EBC). Barring any chemical interaction between the glass/coating, it is unlikely that infiltration will be halted by intrinsic glass crystallization, even with the assumption that there is a thermal gradient in the coating. A 250 μm thick EB-PVD YSZ TBC was nearly completely infiltrated after only 1 h at 1200 °C in contact with an artificial volcanic ash glass [39]. Another study by Mechnich et al. investigated an alternative TBC material, $Gd_2Zr_2O_7$ (GZO), in contact with the artificial volcanic ash and saw slowed infiltration (50 μm after 1 h at 1200 °C) compared to YSZ [41]. This slowing was likely due to the dissolution of the coating in the glass to form newly crystallized reaction products, including an oxyapatite phase, $Ca_2Gd_8(SiO_4)_6O_2$. In the study by Jang et al., the authors investigated the reaction between volcanic ash and a dense EBC material, Yb_2SiO_5, at 1400 °C [40]. Infiltration was halted by the formation of a thin layer of $Yb_2Si_2O_7$. Yb_2SiO_5, in contact with a CMAS composition more closely approximating that of the synthetic sand glass (Table 1; increased CaO content, decreased SiO_2 content), formed a much thicker reaction layer containing Yb apatite, $Ca_2Yb_8(SiO_4)_6O_2$.

TBC materials YSZ and GZO have higher CTEs than the bulk volcanic ash glass, with values of ~11–12 [42,43]. If a TBC is infiltrated and then cooled, strain from a thermal mismatch between the coating and glass can lead to cracking and delamination. EBC materials, rare earth monosilicates (Y_2SiO_5, Yb_2SiO_5), have CTEs ~6–7.5 [6,44,45], which are similar to that of the bulk volcanic ash glass. These materials, however, will have a relatively high thermal mismatch with an underlying SiC-based CMC. SiC has a CTE of about 4.5–5.5 $\times$ 10^{-6} °C^{-6} [6]. Other proposed EBC materials, rare earth (RE) disilicates ($Y_2Si_2O_7$, $Yb_2Si_2O_7$), have CTEs that more closely match that of SiC, ranging ~3.5–5 $\times$ 10^{-6} °C^{-6} [46,47]. The CTE of the bulk glass is greater than that of the CMC/disilicate system and the CTEs of the intrinsically crystallized phases in the glass are considerably so. CTE values (to 1000 °C) for the crystalline products of Fe_3O_4 and Fe_2O_3 are around 9–15 and 9–12 $\times$ 10^{-6} K^{-1} [48], respectively, and 3 and 1.5 $\times$ 10^{-5} K^{-1} for albite and anorthite, respectively [49]. Similar to that expected in a TBC system, CTE mismatch between the glass/crystallized glass and an EBC will likely result in spallation at the coating/glass interface.

In addition to CTE, mechanical properties such as elastic modulus, hardness, and toughness of the coated CMC can be compromised by CMAS attack. The bulk elastic moduli of TBC materials YSZ and GZO are on the order of 200–250 GPa [42,50–52]. However, the typical in-plane modulus for an EB-PVD TBC is about 30 GPa [13]. Upon infiltration, the coating is stiffened, with an E reaching at least that of the glass (75 GPa), and the in-plane compliance of the TBC is degraded [13]. For current EBCs of γ-$Y_2Si_2O_7$ and β-$Yb_2Si_2O_7$, E is around 155 and 168 GPa [53,54], respectively, while Y_2SiO_5 and Yb_2SiO_5 have values of 123 and 149 GPa, respectively [55,56]. The effect of changes in the elastic modulus is likely not as important in EBCs because they are relatively dense compared to TBCs.

Hardness and toughness values for YSZ and GZO are between ~12–14 and 1–2, respectively [50,52]. The hardness of the volcanic ash glass is much lower than these TBC materials, while the toughness is similar. For the Yb- and Y-based silicates, H_V and K_C values are 6.4 and 2.3, respectively, for Yb_2SiO_5 [56], 7.3 and 2.8 for $Yb_2Si_2O_7$ [54], 5.3 and 2.2 for Y_2SiO_5 [55], and 6.2 and 2.1 for $Y_2Si_2O_7$ [53]. Hardness values reported here for the solidified volcanic ash are on the order of those for the RE silicates; more concerning is the apparent decrease in toughness, which could lead to diminished resistance of the coating to fracture due to CMAS infiltration. It is noted that the volcanic ash glass has a greater toughness than that reported for the synthetic and desert sand glasses.

The viscosity of the volcanic ash glass determined experimentally in this study was much lower than predicted by model calculations. Wiesner et al. reported that the FactSage and Fluegel models better represented experimental viscosity values for the synthetic sand glass composition (Table 1) than the Giordano et al. model [35,36]. While the FactSage and Fluegel model predictions more closely matched experimental data for the volcanic ash glass, compared to the Giordano et al. model, they were still higher by about half an order of magnitude. Preliminary Fourier-transform infrared spectroscopy (FTIR) measurements on the volcanic ash glass suggest that H_2O was incorporated into its structure. The presence of water in glass can significantly lower glass viscosity [57], which is a possible explanation for the lower measured viscosity values. It is important to consider the effect of glass bonding with water when modeling CMAS viscosity, especially when considering an actual engine environment wherein high water vapor partial pressures are expected. It is also possible that none of the models investigated are adequate to describe viscosity for this particular composition. Additionally, the valence state of Fe in the glass is uncertain and can vary as a function of composition, temperature and pO_2, and its role in the glass structure. The valence state's effect on glass structure is related to the overall viscosity of the melt. The available iron oxide inputs for the Fluegel and Giordano et al. models are limited to Fe_2O_3 and FeO, respectively. The FactSage model can incorporate both Fe_2O_3 and FeO as input compositions. When substituting FeO for Fe_2O_3 in the volcanic ash glass composition given by Table 1, the predicted viscosity was lower (Table 8), but not to the degree given by experimental data. It is proposed that all Fe^{2+} acts as a glass network modifier while Fe^{3+} acts as a network former when $Fe^{3+}:Fe^{2+} > 1$ [58], explaining the observed changes in calculated viscosity.

Table 8. Model FactSage viscosity (η) values using Fe_2O_3 or FeO as inputs. Using FeO in place of Fe_2O_3, the composition was the same as that for the Giordano et al. model ($5.1CaO–2.5MgO–15.2Al_2O_3–59.9SiO_2–9.3FeO–1.7TiO_2–4.6Na_2O–1.7K_2O$ wt %).

Temperature (°C)	Logη (Pa·s) Using Fe_2O_3	Logη (Pa·s) Using FeO
1500	1.62	1.44
1475	1.74	1.55
1450	1.85	1.66
1425	1.98	1.78
1400	2.11	1.90
1375	2.24	2.03
1350	2.38	2.16
1325	2.52	2.30

The infiltration times calculated for the synthetic sand glass and the volcanic ash glass determined from experimental viscosity values are not very different. The expected time to complete infiltration (determined using Equation (7) and experimental viscosity values; Table 7) is only slightly longer for coatings exposed to the volcanic ash glass (1.6 min at 1410 °C) compared to the synthetic sand glass (0.41 min at 1411 °C). This small difference in infiltration time is likely due to a slightly higher viscosity of the volcanic ash glass resulting in part from a greater SiO_2 and lesser CaO content.

Equation (7) has limitations in its ability to predict the infiltration rate in T/EBCs. The competing effect of crystallization in the determination of infiltration time should be considered for a TBC such as GZO. This coating material dissolves to some degree in contact with CMAS and contributes metal ions to the melt, inducing crystallization of oxyapatite. In addition, this equation is not well suited to predict infiltration in APS-deposited EBCs. APS EBCs are nominally dense and do not contain open channels for CMAS to penetrate (although grain boundaries and pores are susceptible). However, regardless of type, a coating material that reacts quickly with the volcanic ash glass to form newly crystallized phases will likely slow penetration. The low viscosities (Table 7) and sluggish crystallization reported for the volcanic ash glass studied here can lead to fast infiltration via coating channels and defects.

5. Summary and Conclusions

The intrinsic thermal and mechanical properties of a volcanic ash glass were studied using a variety of characterization techniques. It was determined that the glass had a low propensity to crystallize in bulk form, with ≤20 wt % transforming to Fe_3O_4, Fe_2O_3, and/or plagioclase after up to 40 h at temperatures of 900–1200 °C. This was in contrast to a previously investigated synthetic sand glass, which was more easily able to crystallize and formed wollastonite ($CaSiO_3$) and diopside ($Ca(Mg,Al)(Si,Al)_2O_6$). The ability of the volcanic ash glass to crystallize was improved when exposed in powder form.

The melting temperature (T_m), glass transition temperature (T_g), and glass softening temperature (T_d) of the volcanic ash glass were greater than those of sand glass compositions. There was a broad melting range for the volcanic ash glass, determined by DTA; after heating to 1300 °C at 10 °C/min and quenching in air, the glass was mostly amorphous by XRD analysis. The coefficient of thermal expansion (CTE) and Young's modulus (E) of the volcanic ash glass were lower than those reported for sand glasses. Hardness values were similar, but the indentation fracture toughness of the volcanic ash glass was about twice that of the sand glasses, likely due to the presence of some Fe_3O_4 crystallites in the tested sample.

The viscosity of the volcanic ash glass was lower than expected from model predictions. This is unlike that reported for the synthetic sand glass composition, which showed relatively good agreement with Fluegel and FactSage viscosity models. The discrepancies in this study could be due to the difference in the glass composition, or possibly the incorporation of water in the volcanic ash glass structure. Experimental viscosities for the volcanic ash glass were higher than for the synthetic sand glass, in part likely due to a greater SiO_2 and lesser CaO content.

In conclusion, it has been shown that the chemical and mechanical properties of the Eyjafjallajökull volcanic ash CMAS glass can vary significantly from sand-based CMAS glass compositions. In assessing the potential of new coating (T/EBCs) materials for use over a wide range of operating conditions, the variation of CMAS properties with composition must be understood.

Author Contributions: Conceptualization, V.L.W.; methodology, V.L.W. and R.I.W.; validation, V.L.W., J.A.S. and N.P.B.; formal analysis, R.I.W.; investigation, R.I.W.; resources, V.L.W., J.A.S., N.P.B. and E.J.O.; data curation, V.L.W., J.A.S., N.P.B. and R.I.W.; writing—original draft preparation, R.I.W.; writing—review and editing, V.L.W., J.A.S., N.P.B. and E.J.O.; visualization, V.L.W.; supervision, V.L.W., J.A.S. and N.P.B.; project administration, V.L.W.; funding acquisition, V.L.W. and E.J.O. All authors have read and agree to the published version of the manuscript.

Funding: This research was supported by NASA's Transformative Tools and Technologies (TTT) Project within the Transformative Aeronautics Concept Program (TCAP) and the Lewis' Educational and Research Collaborative Internship Project (LERCIP) at NASA Glenn Research Center.

Acknowledgments: The authors are grateful to Gustavo Costa, Bryan Harder, Nathan Jacobson, Dereck Johnson, and Richard Rogers for training and helpful discussion, and Marie Kløve Keiding, Magnus Gudmundsson, and Tinna Jónsdóttir for assistance in obtaining volcanic ash from the Eyjafjallajökull eruption used in this study.

Conflicts of Interest: The authors declare no conflict of interest.

References

1. De Wet, D.; Taylor, R.; Stott, F. Corrosion mechanisms of ZrO_2-Y_2O_3 thermal barrier coatings in the presence of molten middle-east sand. *J. Phys. IV* **1993**, *3*, 655–663. [CrossRef]

2. Stott, F.H.; De Wet, D.; Taylor, R. Degradation of thermal barrier coatings at very high temperature. *MRS Bull.* **1994**, *19*, 46–49. [CrossRef]

3. Clarke, D.R.; Oechsner, M.; Padture, N.P. Thermal-barrier coatings for more efficient gas turbine engines. *MRS Bull.* **2012**, *37*, 891–898. [CrossRef]

4. Ohnabe, H.; Masaki, S.; Onozuka, M.; Miyahara, K.; Sasa, T. Potential application of ceramic matrix composites to aero-engine components. *Compos. Part A* **1999**, *30*, 489–496. [CrossRef]

5. Lee, K.N. Environmental Barrier Coatings for SiC_f/SiC. In *Ceramic Matrix Composites: Materials, Modeling and Technology*; Bansal, N.P., Lamon, J., Eds.; John Wiley and Sons: Hoboken, NJ, USA, 2014; pp. 430–451.

6. Lee, K.N.; Fox, D.S.; Bansal, N.P. Rare earth silicate environmental barrier coatings for SiC/SiC composites and Si_3N_4 ceramics. *J. Eur. Ceram. Soc.* **2005**, *25*, 1705–1715. [CrossRef]

7. Borom, M.P.; Johnson, C.A.; Peluso, L.A. Role of environment deposits and operating surface temperature in spallation of air plasma sprayed thermal barrier coatings. *Surf. Coat. Technol.* **1996**, *86*, 116–126. [CrossRef]

8. Krämer, S.; Yang, J.; Levi, C.G. Infiltration-inhibiting reaction of gadolinium zirconate thermal barrier coatings with CMAS melts. *J. Am. Ceram. Soc.* **2008**, *91*, 576–583. [CrossRef]

9. Poerschke, D.L.; Jackson, R.W.; Levi, C.G. Silicate deposit degradation of engineered coatings in gas turbines: Progress toward models and materials solutions. *Annu. Rev. Mater. Res.* **2017**, *47*, 297–330. [CrossRef]

10. Stolzenburg, F.; Johnson, M.T.; Lee, K.N.; Jacobson, N.S.; Faber, K.T. The interaction of calcium magnesium-aluminosilicate with ytterbium silicate environmental barrier materials. *Surf. Coat. Technol.* **2015**, *284*, 44–50. [CrossRef]
11. Zhao, H.; Richards, B.T.; Levi, C.G.; Wadley, H.N.G. Molten silicate reactions with plasma sprayed ytterbium silicate coatings. *Surf. Coat. Technol.* **2016**, *288*, 151–162. [CrossRef]
12. Ahlborg, N.L.; Zhu, D. Calcium-magnesium aluminosilicate (CMAS) reactions and degradation mechanisms of advanced environmental barrier coatings. *Surf. Coat. Technol.* **2013**, *237*, 79–87. [CrossRef]
13. Wiesner, V.L.; Harder, B.J.; Bansal, N.P. High-temperature interactions of desert sand CMAS glass with yttrium disilicate environmental barrier coating material. *Ceram. Int.* **2018**, *44*, 22738–22743. [CrossRef]
14. Levi, C.G.; Hutchinson, J.W.; Vidal-Sétif, M.-H.; Johnson, C.A. Environmental degradation of thermal barrier coatings by molten deposits. *MRS Bull.* **2012**, *37*, 932–941. [CrossRef]
15. Zaleski, E.M.; Ensslen, C.; Levi, C.G. Melting and crystallization of silicate systems relevant to thermal barrier coating damage. *J. Am. Ceram. Soc.* **2015**, *98*, 1642–1649. [CrossRef]
16. Delmelle, P.; Lambert, M.; Dufrêne, Y.; Gerin, P.; Óskarsson, N. Gas/aerosol-ash interaction in volcanic plumes: New insights from surface analyses of fine ash particles. *Earth Planet. Sci. Lett.* **2007**, *259*, 159–170. [CrossRef]
17. Wiesner, V.L.; Bansal, N.P. Crystallization kinetics of calcium-magnesium aluminosilicate (CMAS) glass. *Surf. Coat. Technol.* **2014**, *259*, 608–615. [CrossRef]
18. Wiesner, V.L.; Bansal, N.P. Mechanical and thermal properties of calcium magnesium aluminosilicate (CMAS) glass. *J. Eur. Ceram. Soc.* **2015**, *35*, 2907–2914. [CrossRef]
19. Bansal, N.P.; Choi, S.R. Properties of CMAS glass from desert sand. *Ceram. Int.* **2015**, *41*, 3901–3909. [CrossRef]
20. *ASTM C1259-14 Standard Test Method for Dynamic Young's Modulus, Shear Modulus, and Poisson's Ratio for Advanced Ceramics by Impulse Excitation of Vibration*; ASTM International: Conshohocken, PA, USA, 2014.
21. *ASTM C1327-08 Standard Test Method for Vickers Indentation Hardness of Advanced Ceramics*; ASTM International: Conshohocken, PA, USA, 2008.
22. Lamkin, M.A.; Riley, F.L.; Fordham, R.J. Oxygen mobility in silicon dioxide and silicate glasses: A review. *J. Eur. Ceram. Soc.* **1992**, *10*, 347–367. [CrossRef]
23. Monazam, E.R.; Breault, R.W.; Siriwardane, R. Kinetics of magnetite (Fe_3O_4) oxidation to hematite (Fe_2O_3) in air for chemical looping combustion. *Ind. Eng. Chem. Res.* **2014**, *53*, 13320–13328. [CrossRef]
24. Forsmo, S.P.E.; Forsmo, S.-E.; Samskog, P.-O.; Björkman, B.M.T. Mechanisms in oxidation and sintering of magnetite iron ore green pellets. *Powder Technol.* **2008**, *183*, 247–259. [CrossRef]
25. Bale, C.W.; Bélisle, E.; Chartrand, P.; Decterov, S.A.; Eriksson, G.; Hack, K.; Jung, I.-H.; Kang, Y.B.; Melançon, J.; Pelton, A.D.; et al. FactSage thermochemical software and databases-recent developments. *Calphad* **2009**, *33*, 295–311. [CrossRef]
26. Miyoshi, T.; Sagawa, N.; Sassa, T. Study on fracture toughness evaluation for structural ceramics. *Trans. Jpn. Soc. Mech. Eng.* **1985**, *51*, 2487–2489. [CrossRef]
27. Marshall, D.; Evans, A. Reply to "Comment on 'Elastic/plastic indentation damage in ceramics: The median/radial crack system'". *J. Am. Ceram. Soc.* **1981**, *64*, C182–C183.
28. Anstis, G.; Chantikul, P.; Lawn, B.R.; Marshall, R. A critical evaluation of indentation techniques for measuring fracture toughness: I. Direct crack measurements. *J. Am. Ceram. Soc.* **1981**, *64*, 533–538. [CrossRef]
29. Salem, J.; Jenkins, M. Applying ASTM International C1421 to glasses and optical ceramics. In *Proceedings of the 42nd International Conference on Advanced Ceramics and Composites, Ceramic Engineering and Science Proceedings*; Salem, J., Koch, D., Mechnich, P., Kusnezoff, M., Bansal, N.P., LaSalvia, J.C., Balaya, P., Fu, Z., Ohii, T., Wiesner, V., et al., Eds.; John Wiley & Sons: Hoboken, NJ, USA, 2019; Volume 39, pp. 27–40.
30. Giordano, D.; Russell, J.K.; Dingwell, D.B. Viscosity of magmatic liquids: A model. *Earth Planet. Sci. Lett.* **2008**, *271*, 123–134. [CrossRef]
31. Vogel, D.H. Temperaturabhängigkeitsgesetz der Viskosität von Flüssigkeiten. *Phys. Z.* **1921**, *22*, 645–646.
32. Fulcher, G.S. Analysis of recent measurements of the viscosity of glasses. *J. Am. Ceram. Soc.* **1925**, *8*, 339–355. [CrossRef]
33. Tammann, G.; Hesse, W. Die abhängigkeit der viskosität von der temperatur bei unterkühlten flüssigkeiten. *Z. Anorgan. Allg. Chem.* **1926**, *156*, 245–257. [CrossRef]

34. Fluegel, A. Glass viscosity calculation based on a global statistical modeling approach. *Glass Technol. Eur. J. Glass Sci. Technol. Part A* **2007**, *48*, 13–30.

35. Wiesner, V.L.; Vempati, U.K.; Bansal, N.P. High temperature viscosity of calcium magnesium aluminosilicate glass from synthetic sand. *Scr. Mater.* **2016**, *124*, 189–192. [CrossRef]

36. Wiesner, V.L.; Vempati, U.K.; Bansal, N.P. Corrigendum to "High temperature viscosity of calcium-magnesium-aluminosilicate glass from synthetic sand" [Scripta Mater. 2016, 124, 189–192]. *Scr. Mater.* **2017**, *130*, 298. [CrossRef]

37. Krämer, S.; Yang, J.; Levi, C.G.; Johnson, C.A. Thermochemical interaction of thermal barrier coatings with molten CaO-MgO-Al$_2$O$_3$-SiO$_2$ (CMAS) deposits. *J. Am. Ceram. Soc.* **2006**, *89*, 3167–3175. [CrossRef]

38. Kucuk, A.; Clare, A.; Jones, L. An estimation of the surface tension for silicate glass melts at 1400 °C using statistical analysis. *Glass Technol.* **1999**, *40*, 149–153.

39. Mechnich, P.; Braue, W.; Schulz, U. High-temperature corrosion of EB-PVD yttria partially stabilized zirconia thermal barrier coatings with an artificial volcanic ash overlay. *J. Am. Ceram. Soc.* **2011**, *94*, 925–931. [CrossRef]

40. Jang, B.-K.; Feng, F.-J.; Suzuta, K.; Tanaka, H.; Matsushita, Y.; Lee, K.-S.; Ueno, S. Corrosion behavior of volcanic ash and calcium magnesium aluminosilicate on Yb$_2$SiO$_5$ environmental barrier coatings. *J. Ceram. Soc. Jpn.* **2017**, *125*, 326–332. [CrossRef]

41. Mechnich, P.; Braue, W. Volcanic ash-induced decomposition of EB-PVD Gd$_2$Zr$_2$O$_7$ thermal barrier coatings to Gd-oxyapatite, zircon, and Gd,Fe-zirconolite. *J. Am. Ceram. Soc.* **2013**, *96*, 1958–1965. [CrossRef]

42. Evans, A.G.; Mumm, D.R.; Hutchinson, J.W.; Meier, G.H.; Pettit, F.S. Mechanisms controlling the durability of thermal barrier coatings. *Prog. Mater. Sci.* **2001**, *46*, 505–553. [CrossRef]

43. Touloukian, Y.S.; Kirby, R.K.; Taylor, R.E.; Lee, T.Y.R. *Thermal Expansion-Non Metallic Solids*; IFI/Plenum: New York, NY, USA, 1977; Volume 13.

44. Fukuda, K.; Matsubara, H. Anisotropic thermal expansion in yttrium silicate. *J. Mater. Res.* **2003**, *18*, 1715–1722. [CrossRef]

45. NETL. *The Gas Turbine Handbook*; United States Department of Energy (DOE): Morgantown, WV, USA, 2006.

46. Sun, Z.; Zhou, Y.; Wang, J.; Li, M. Thermal properties and thermal shock resistance of γ- Y$_2$Si$_2$O$_7$. *J. Am. Ceram. Soc.* **2008**, *91*, 2623–2629. [CrossRef]

47. Fernández-Carrión, A.J.; Allix, M.; Becerro, A.I. Thermal expansion of rare-earth pyrosilicates. *J. Am. Ceram. Soc.* **2013**, *96*, 2298–2305. [CrossRef]

48. Takeda, M.; Onishi, T.; Nakakubo, S.; Fujimoto, S. Physical properties of iron-oxide scales on Si containing steels at high temperature. *Mater. Trans.* **2009**, *50*, 2242–2246. [CrossRef]

49. Tribaudino, M.; Angel, R.J.; Cámara, F.; Nestola, F.; Pasqual, D.; Margiolaki, I. Thermal expansion of plagioclase feldspars. *Mineral. Petrol.* **2010**, *160*, 899–908. [CrossRef]

50. Vassen, R.; Cao, X.; Tietz, F.; Basu, D.; Stöver, D. Zirconates as new materials for thermal barrier coatings. *J. Am. Ceram. Soc.* **2000**, *83*, 2023–2028. [CrossRef]

51. Van Dijk, M.P.; de Vries, K.J.; Burggraaf, A.J. Oxygen ion and mixed conductivity in compounds with the fluorite and pyrochlore structure. *Solid State Ion.* **1983**, *9–10*, 913–919. [CrossRef]

52. Schmitt, M.P.; Stokes, J.L.; Gorin, B.L.; Rai, A.K.; Zhu, D.; Eden, T.J.; Wolfe, D.E. Effect of Gd content on mechanical properties and erosion durability of sub-stoichiometric Gd$_2$Zr$_2$O$_7$. *Surf. Coat. Technol.* **2017**, *313*, 177–183. [CrossRef]

53. Sun, Z.; Zhou, Y.; Wang, J.; Li, M. γ-Y$_2$Si$_2$O$_7$, a machinable silicate ceramic: Mechanical properties and machinability. *J. Am. Ceram. Soc.* **2007**, *90*, 2535–2541. [CrossRef]

54. Zhou, Y.-C.; Zhao, C.; Wang, F.; Sun, Y.-J.; Zheng, L.-Y.; Wang, X.-H. Theoretical prediction and experimental investigation on the thermal and mechanical properties of bulk β- Yb$_2$Si$_2$O$_7$. *J. Am. Ceram. Soc.* **2013**, *96*, 3891–3900. [CrossRef]

55. Sun, Z.; Wang, J.; Li, M.; Zhou, Y. Mechanical properties and damage tolerance of Y$_2$SiO$_5$. *J. Eur. Ceram. Soc.* **2008**, *28*, 2895–2901. [CrossRef]

56. Lu, M.-H.; Xiang, H.-M.; Feng, Z.-H.; Wang, X.-Y.; Zhou, Y.-C. Mechanical and thermal properties of Yb$_2$SiO$_5$: A promising material for T/EBCs applications. *J. Am. Ceram. Soc.* **2016**, *99*, 1401–1411. [CrossRef]

57. Stolper, E. Water in silicate glasses: An infrared spectroscopic study. *Contrib. Mineral. Petrol.* **1982**, *81*, 1–17. [CrossRef]
58. Cook, G.B.; Cooper, R.F. Chemical diffusion and crystalline nucleation during oxidation of ferrous iron-bearing magnesium aluminosilicate glass. *J. Non Cryst. Solids* **1990**, *120*, 207–222. [CrossRef]

Article

Flow Kinetics of Molten Silicates through Thermal Barrier Coating: A Numerical Study

Mohammad Rizviul Kabir [1,*], Anil Kumar Sirigiri [2], Ravisankar Naraparaju [2] and Uwe Schulz [2]

[1] Experimental and Numerical Methods, Institute of Materials Research, German Aerospace Center (DLR), Linder Höhe, 51147 Cologne, Germany

[2] High Temperature and Functional Coatings, Institute of Materials Research, German Aerospace Center (DLR), Linder Höhe, 51147 Cologne, Germany; anilkumars135@gmail.com (A.K.S.); Ravisankar.Naraparaju@dlr.de (R.N.); Uwe.Schulz@dlr.de (U.S.)

* Correspondence: mohammad-rizviul.kabir@dlr.de; Tel.: +49-2203-601-2481

Received: 16 April 2019; Accepted: 21 May 2019; Published: 23 May 2019

Abstract: Infiltration of molten calcium–magnesium–alumina–silicates (CMAS) through thermal barrier coatings (TBCs) causes structural degradation of TBC layers. The infiltration kinetics can be altered by careful tailoring of the electron beam physical vapor deposition (EB-PVD) microstructure such as feather arm lengths and inter-columnar gaps, etc. Morphology of the feathery columns and their inherent porosities directly influences the infiltration kinetics of molten CMAS. To understand the influence of columnar morphology on the kinetics of the CAMS flow, a finite element based parametric model was developed for describing a variety of EB-PVD top coat microstructures. A detailed numerical study was performed considering fluid-solid interactions (FSI) between the CMAS and TBC top coat (TC). The CMAS flow characteristics through these microstructures were assessed quantitatively and qualitatively. Finally, correlations between the morphological parameters and CMAS flow kinetics were established. It was shown that the rate of longitudinal and lateral infiltration could be minimized by reducing the gap between columns and increasing the length of the feather arms. The results also show that the microstructures with long feather arms having a lower lateral inclination decrease the CMAS infiltration rate, therefore, reduce the CMAS infiltration depth. The analyses allow the identification of key morphological features that are important for mitigating the CMAS infiltration.

Keywords: TBCs; CMAS; infiltration; microstructure; modeling; finite element

---

1. Introduction

Hot section engine components of gas turbines are insulated with thermal barrier coatings (TBCs) to protect them against excessive heat. The TBC layers fabricated by electron beam physical vapor deposition (EB-PVD) show superior structural stability and long-life compared to the TBCs produced by the atmospheric plasma spray (APS) method due to their feathery columnar microstructure [1–6]. However, state-of-the art standard 7YSZ (7 wt.% yttria stabilized zirconia) coatings are vulnerable against environmental attacks, i.e., sand, volcanic ash, etc., that comes from dusty air intake during engine operation. These silicate-based particles melt under high engine temperature forming calcium–magnesium–alumina–silicates (CMAS), deposit on coated parts, and infiltrates through the porous channels of the TBC layers [7,8]. The contamination of CMAS eventually degrades the structural properties of TBCs, causing spallation failure under thermal cycles [9–13].

In past years, considerable attempts were made on the improvement of TBC systems aiming at reducing the infiltration and contamination by molten silicates [14–19]. Several methodologies were suggested to enforce the mitigation mechanisms against CMAS, which fall under two broad categories. In the first approach, new TBC compositions were proposed that react with CMAS, and recrystallize into new stable phases, which restrict further CMAS infiltration [20–23]. A long-term resistance against the CMAS flow cannot be guaranteed due to the continuous reaction and recession of the TBC material. In the second approach, microstructures of the top coat were custom-tailored for distributed porosities, through which the deposition of CMAS occurs locally over the top coat. After solidification of CMAS inside these pores, further infiltration was restricted, thus, the total infiltration depth was reduced.

The latter approach was systematically analyzed in some previous experimental works [24–26], where the CMAS infiltration behavior was compared for different top coat microstructures. It was found out that by using the optimized columnar microstructure the CMAS infiltration kinetics could be changed, where infiltration mainly proceed due to the capillary forces. The results suggest that custom-tailored well-defined porosities of the TBC microstructure may potentially impede the infiltration of CMAS. However, the knowledge provided by the previous works is not sufficient for identifying such an optimized morphology of TBCs. A systematic analysis to understand the microstructure dependent infiltration behavior is highly required. This goal can be achieved by conducting additional experiments with a number of columnar structures obtained via EB-PVD, which involves many trial-and-error modifications of the top coat morphology. Additionally, advanced numerical analyses can be suggested to predict the microstructural influence on infiltration kinetics, which will obviously reduce the experimental efforts, and additionally, will provide an in-depth understanding of the infiltration processes.

In past decades, the numerical investigations of TBCs were performed focusing mostly on the problems of structural stability, life-cycle reduction, and failure [27–30]. The effects of deposited CMAS were investigated in terms of structural property degradation. The CMAS infiltration depth was assumed based on the experimental observation. As the infiltration depth is dependent on the morphological factors of the TBC, it is desirable to develop models that firstly describe the infiltration kinetics, and secondly, predict the infiltration depth considering the morphological obstacles.

In one of the recent works by the authors [24], the infiltration depth of CMAS through different porous EB-PVD top coats was modeled analytically, where the longitudinal porous channels were assumed as open or concentric pipes, through which CMAS infiltrated under the action of capillary forces. Microstructure obstacles imposed on the CMAS flow was considered by a parameter "tortuosity factor". The model predicted the microstructure dependency of the infiltration depth by fitting tortuosity factors for different morphologies. A good approximation of the infiltration time compared to depth was obtained which closely predicts the experimental observation [31]. However, interest on direct microstructure modeling incorporating the flow behavior is growing tremendously, as this approach allows describing realistic models based on actual microstructural statistics, and also allows fluid structure interactions during liquid flow. Thus, no "tortuosity" assumptions are required to be imposed on the moving liquid.

Microstructure based models also give an opportunity to analyze a number of equivalent real microstructure numerically, which give an easy access to the details of molten CMAS flow kinetics that are not easily accessible by experimental means. The know-how may provide qualified information for custom-tailoring the morphology of the EB-PVD top coat that potentially acts as a barrier to the CMAS infiltration.

So far, no detailed modeling approach was found that captures microstructure-sensitivity of the CMAS flow kinetics, however, the demand of an advanced modeling framework is increasing. In the present work, an effort has been made to establish a preliminary computational and numerical framework for such analysis. The following main objectives are focused:

- Developing the computational modeling approach for the CMAS infiltration simulation: The main focus is to develop parametric geometric models of EB-PVD top coats based on microstructural statistics. For the infiltration simulation a coupled Lagrangian–Eulerian modeling framework will be adopted.
- Quantitative and qualitative assessment of the CMAS flow kinetics: Here, a detailed comparative study of the CMAS flow kinetics for different morphologies will be performed. Finally, a clear

insight into the flow behavior with respect to the individual morphological parameter will be obtained.

2. Materials and Methods

2.1. Material and Microstructure

Thermal barrier coating (TBC) systems are generally comprised of three layers: (a) a bond coat layer (BC) of a Ni-rich nickel aluminide or MCrAlY on Ni-based superalloy substrate; (b) thermally grown alumina oxide layer (TGO); and (c) a top coat (TC) typically of 7YSZ. Figure 1 shows an EB-PVD fabricated 7YSZ-TBC on a Ni-based superalloy component. The top coat consists of a very fine feathery columnar microstructure. The thickness of the columns decreases from top to bottom, and the size of the porous channel reduces accordingly. At the bottom, very dense and fine thin columns are generated. Very fine nanopores can be identified at this zone.

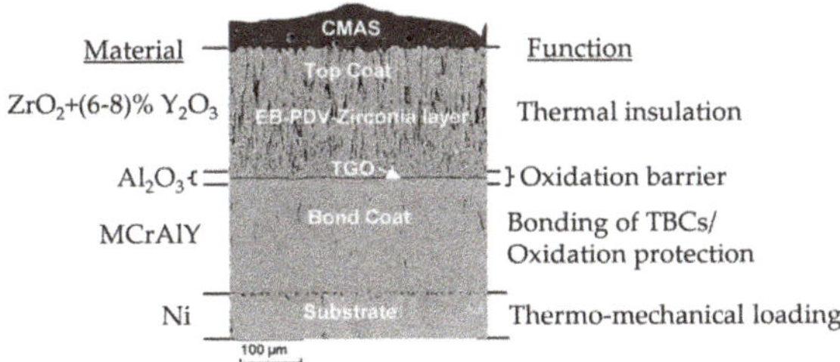

Figure 1. Cross-section of electron beam physical vapor deposition (EB-PVD) fabricated thermal barrier coating (TBC) showing layered structure with deposited calcium–magnesium–alumina–silicates (CMAS) on top.

Variation of top coat microstructures can be obtained by custom tailoring the columnar morphology with controlled EB-PVD process parameters [25,32]. Several tailored microstructures with different feathery columnar lengths are shown in Figure 2. The morphological details can be identified from the top and cross-sectional views.

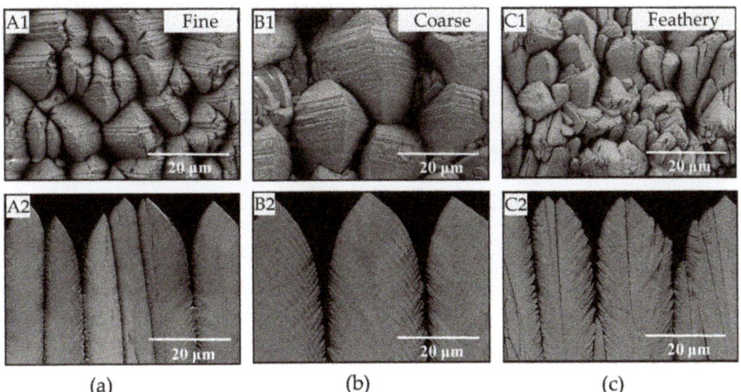

Figure 2. Tailored TBC microstructures fabricated by EB-PVD. Top view (upper A1, B1 and C1) and cross-sectional views (bottom A2, B2 and C2) are shown for: (**a**) Fine columnar microstructure; (**b**) coarse columnar microstructure; and (**c**) columnar microstructure with feathery arms (high porosity). Reprinted with permission from [25]; Copyright 2006 Elsevier.

Here, only three exemplary morphologies, named Fine, Course, and Feathery are shown, which vary with respect to the columnar microstructure. The cross-sectional views A2, B2, and C2 show numerous single crystalline columns and nano-sized feather arms along the complete column's periphery. The top views A1, B1, and C1 show different columnar tips, which also vary with respect to size and shape. The overall porosity of the top coat can be attributed to the gaps between columns, feathery arms, and gaps in the feather tips. Due to the morphological variations in Fine, Course, and Feathery columns, the overall porosity has been altered.

2.2. Microstructure Modeling of EB-PVD Top Coat

2.2.1. Microstructure Idealization

Geometries of the top coat columns and feathers are highly complex. Moreover, the microstructure of each column varies slightly with each other. For modeling purposes, the microstructure variations need to be idealized in a statistical manner providing that the main structural features remain consistent. Therefore, a statistically representative microstructure unit cell (MUC) needs to be defined with much flexibility to incorporate necessary geometrical details as found in different types of EB-PVD top coats. Details of the MUC are given in Section 2.2.2.

To generate such a MUC, the microstructure parameters are identified from a high magnification SEM micrograph of an EB-PVD TBC column as shown in Figure 3a. The total column diameter is D_{TC}, and the columnar diameter excluding the feather arm length is the reduced column diameter, D_C, as shown in the figure. The "reduced column diameter" will be termed simply as "column diameter" in the following sections unless otherwise stated. Similarly, for the sake of brevity, the reduced column radius, which is the half of the reduced column diameter, will be termed as "column radius".

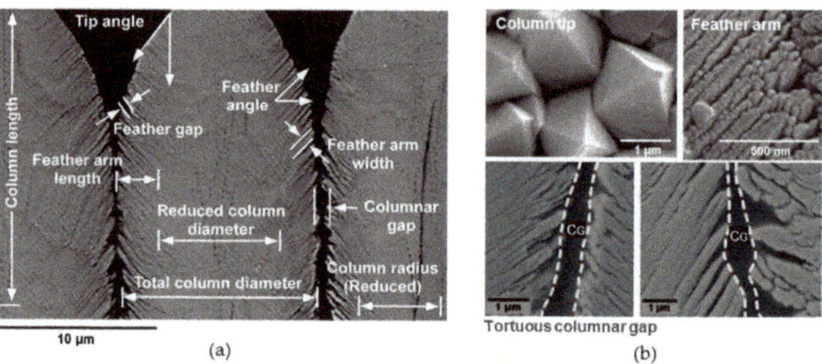

Figure 3. SEM image of a cross-sectional view of an EB-PVD 7YSZ column. The microstructures are shown for: (**a**) columnar geometry of the top coat (TC) with microstructural parameters identified for modeling purposes; and (**b**) closer look of the top view of pyramidal column tip, side view of feathers with irregular feather arrangements (top), and tortuous columnar gaps due to irregular feather arms (bottom). Reprinted with permission from [25]; Copyright 2006 Elsevier.

Other morphological parameters are total length of the column, the column length (L_C), and gap between two columns, the columnar gap (G_C). The thickness of each feather arm was named as the feather arm width or feather thickness (T_F), the horizontal length of the feather arm along the column width is the feather arm length (L_F), and the gap between individual feather arms is considered as void between feather arms (V_F), and will be termed shortly the feather gap. Further, the inclination of feather arms with horizontal axis was considered as the feather inclination angle (θ_F), the vertical angle of the pyramidal tip is the feather tip angle (θ_T).

Figure 3b shows additional microstructural details with morphological variations of columnar tips, feather arms, and tortuosity due to feather arms. The pyramidal tip angle is around 30°–45°.

The facets of the tips depict staggered arrangements causing gaps between pyramidal tips. The feather arms are inclined at a 40°–60° angle. The length of the feather arm depends on this angle of inclination.

Bottom pictures of the Figure 3b show two cross-sectional views of the tortuous columnar gaps due to uneven staggered arrangements of feathery arms. The feather arms and feather tips are irregular in shape and size along this columnar gap, i.e., they vary at left and right columnar edge as well as from top to bottom of a column. The degree of irregularities causes high or low tortuosity along columnar gaps. The columnar gaps may show almost regular width along the column length with a low tortuosity over the column height, or they might be highly irregular with high tortuosity contour along the column length. It must be noted that all microstructural features are three-dimensional, and they may vary along the circumference of each column considerably. The microstructures analyzed here are two-dimensional cuts; however, several orientations of cuts can be taken into account.

From the geometrical details stated above, the most relevant morphological parameters influencing the CMAS flow behavior were identified as illustrated in Figure 4a. A python code was written to handle these variables in a parametric way for generating a number of synthetic (virtual) columnar microstructures resembling real EB-PVD columnar structures. Several exemplary models generated by this code are shown in Figure 4b. A number of possible morphological variations can be obtained by cross-mixing these microstructure parameters.

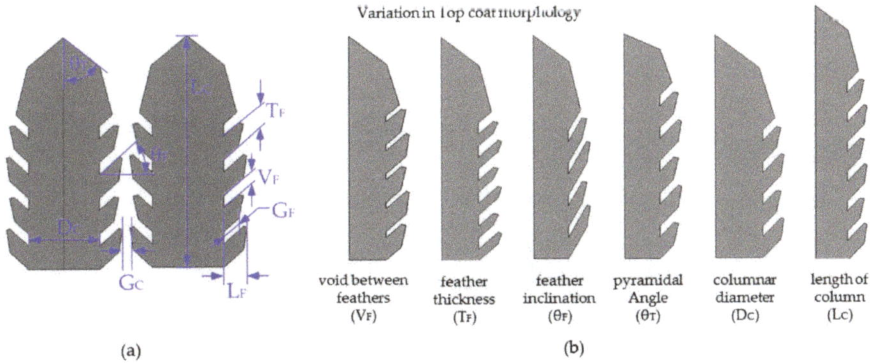

Figure 4. Model representation of the EB-PVD columnar microstructure idealized from real microstructural SEM images: (**a**) definition of morphological parameters for computer generated parametric model of EB-PVD columns; and (**b**) automated model generation using morphological parameter variations. Column structures with different feather geometries were obtained.

In these models, the feather arm length (L_F) can be made short or long with sharp or rounded feather tips to obtain irregularities in feather arrangements as observed in the microstructure picture of Figure 3a,b. A sharp feather tip was defined by an angular cut at the feather arm. The top of the column with pyramidal tip was idealized as a triangular shape keeping the width of the pyramid equal to the columnar diameter. According to the microstructure picture, the pyramidal width increases gradually from top to bottom. This geometry was obtained by defining two gradual angles along the pyramidal edges. The width of the pyramid bottom can be shortened to obtain a large CMAS entry area over the columnar gap and enlarged for obtaining a smaller CMAS entry area.

In this parametric model, the feather arm angle is taken either 40° or 60°. It should be noted that due to trigonometric functions in the parametric model, a higher feather inclination angle results in a thinner feather gap, and a lower inclination angle yields a wider father gap. In reality, the feather gaps are arbitrary, and are not directly correlated with feather inclination angles.

Total porosity of the model can be calculated adding the columnar gaps and voids between feather arms (feather gaps).

2.2.2. Microstructure Unit Cell (MUC) Model

The columnar microstructural setting of the top coat was modeled by a periodic microstructure unit cell (MUC) consisting of symmetric parts of the columnar sub-structures arranged side-by-side, facing each other along the feathery arms (Figure 5). It was assumed that each microstructure is symmetric along the mid-axis of a column. This assumption conforms with our observation that the real EB-PVD TBC microstructure is also nearly symmetric at least along the two main directions, i.e., parallel to the rotational axis and perpendicular to it. For a 2D-simplification, only two EB-PVD columns having a particular columnar gap, C_G, were considered as a statistical feathery unit of columns, as shown in Figure 5a. The 2D-MUC model can be arranged periodically in X-direction to obtain a complete EB-PVD top coat layer. In Z-direction, the columns are assumed to be infinite. For a 3D-simplification, four 1/8th symmetric parts of the feathery columns can be re-arranged in both X- and Z-directions, keeping the symmetry plane at the boundaries, and with a face-to-face setting of the feather arms separated by a columnar gap, C_G. A more realistic EB-PVD top coat columnar microstructure can be obtained by this 3D-MUC model taking idealized symmetric microstructure in both X- and Z-directions, as shown in Figure 5c. In 3D-MUC, the tortuosity in the feathery structures can be modeled taking further geometrical complexity into account. For the present work, the investigations were done on 2D-MUCs only to obtain fundamental understanding of the flow behavior, however, advance analyses with 3D-MUC models are open for the future.

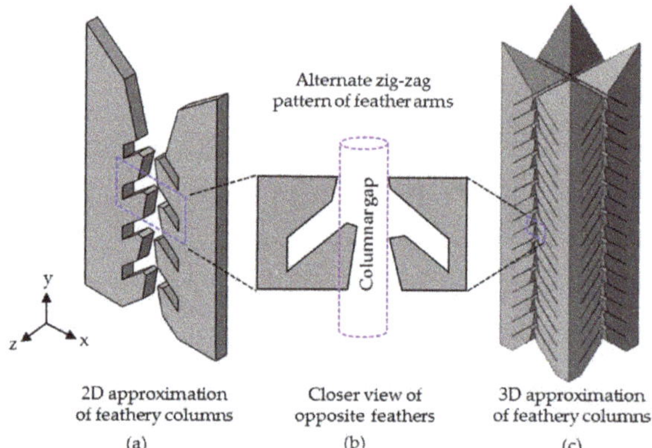

Figure 5. Microstructure unit cell (MUC) for infiltration analysis: (**a**) a two-dimensional (2D) representation of feathery columns with sideway periodic arrangements; (**b**) fine details of the feather tips idealized form real tortuous feathery structures; and (**c**) a three-dimensional (3D) representation of MUC consisting of four one-eighth feathery columns, arranged with alternate zigzag patterns of feathery structures.

For modeling simplicity, the columnar gap was assumed to be vertically straight, and has a constant radius over the length (Figure 3b). The tortuosity along the columnar gap was obtained by defining a particular arrangement of feather arm and feather tip as described in the following.

At first, the feathery arm patterns with non-uniformities were idealized as periodic irregularities, as can be seen in Figure 5a,c along the columnar gaps. A single set of opposite feathers is shown in Figure 5b with enlargement. Here, the feather-tips of the left column were made slightly longer and the feather-tips of the right column were made shorter such that the tips of the longer feathers extend slightly into the columnar gap, and the tips of the shorter feathers stay at the columnar gap. In the left column, a periodic arrangement of longer and shorter feather-arms was done over the length of columnar gaps (from top to bottom) to obtain a realistic irregular (tortuous) feathery microstructure. For the right column, the arrangement is with an opposite order, one feather was made shorter, and the next was longer, and they are successively arranged along top to bottom. Note that the feathery structures of the opposite columns are not the mirror image along the columnar gap, which is further illustrated in the figures of the following sections.

For the numerical analysis of infiltration kinetics, only a 2D-MUC was used. The geometries of the top coat columns were varied in the MUC in order to obtain different types of EB-PVD morphologies.

2.3. Numerical Approach

2.3.1. Modeling Approach for Molten CMAS Flow

The molten CMAS flow through EB-PVD top coats involves multi-physics Fluid (Liquid)-Solid Interactions (FSI), which were modeled using a coupled Eulerian–Lagrangian (CEL) approach available in the commercial software ABAQUS [33]. The top coat was defined as a solid structure in the Lagrangian framework allowing structural deformation and rotation under the elastoplastic material behavior. On the other hand, the molten CMAS was assumed to be a highly dense, compressible, viscous fluid. To model the severe deformation and distortion of the molten CMAS, an Eulerian approach was adopted. Further, the flow of CMAS was assumed to be Newtonian, as the infiltration of the CMAS through TBC occurs only by capillary actions in the presence of the low strain-rate, i.e., strain rate effects on the viscous flow was neglected. Only the temperature dependency on viscosity was considered.

To capture multi-physics interactions along the solid-liquid domain boundaries (here, the domain between top coat and CMAS), the free surface evolution of the liquid CMAS flow was modeled using a volume-of-fluid (VOF) method. In this approach, the prescribed liquid material is contained within a defined domain of Eulerian elements bounded by a large domain. The amount of liquid material that is filled or emptied in an Eulerian element is defined by the Eulerian Volume Fraction (EVF) for that element. An element is fully filled with a material if EVF = 1.0, and this element is empty (void element) if EVF = 0. EVF ranges within 0.25, 0.5, and 0.75 for partially filled material. The free surface of a liquid flow is approximated from the boundaries of fully filled, partially filled or voided elements. During the material flow, the free surface of the moving material is tracked, and approximated as a continuous flow-front.

Interactions between Lagrangian solids, and Eulerian liquids were modeled by defining particular contact interactions among these two domain boundaries. During the liquid flow, the evolving Eulerian surface exerts pressure on the solid surfaces according to the flow density and pre-defined friction behavior, and both the liquid-solid interface share a common deformation behavior. However, this approach requires intensive computational power. For the present work, some computational efforts were reduced by assuming the EB-PVD columnar structures as rigid bodies with no translational and rotational degree of freedom. This assumption was acceptable for the proposed analysis, as the goal was primarily focused on the infiltration kinetics of the molten CMAS, but not the deformation and stress evolution in the TBCs.

In Figure 6, the microstructure unit cell (MUC) was re-arranged for FSI modeling in a CEL framework. To define the Eulerian CMAS domain, an isosceles trapezoidal shape of the CMAS domain was identified according to the observation of the real microstructure (Figure 6a). Accordingly, an isosceles trapezoid sub-domain was defined over the void area of the MUC model as shown in Figure 6b. This CMAS sub-domain was assumed to be in the molten state, which has already been moved (or infiltrated) up to the column tips. To capture the primary effects of morphological influences on the infiltration behavior, it is sufficient to model the CMAS flow through the columnar gaps taking different feather morphologies.

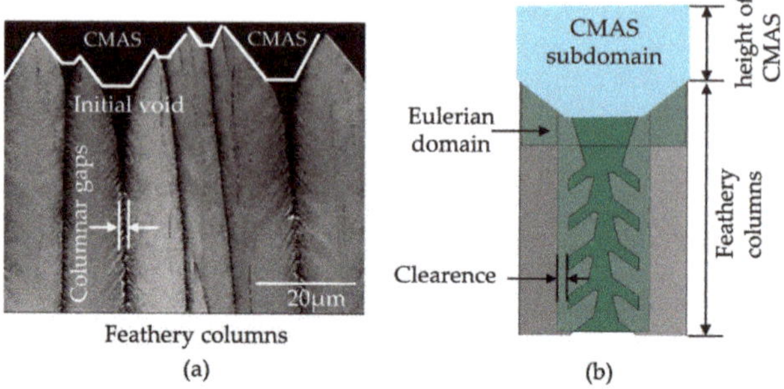

Figure 6. Identification of the CMAS initial position for modeling: (**a**) trapezoidal shape of the CMAS domain is isolated at the top of the columns; and (**b**) microstructure unit cell (MUC) with CMAS sub-domain and Eulerian domains for material flow. Reprinted with permission from [25]; Copyright 2006 Elsevier.

For this simulation, an additional Eulerian domain with void elements need to be defined such that the domain occupies the columnar as well as feathery gaps, and it extends slightly beyond the Lagrangian domains by a small clearance. This clearance was required to ensure proper contacts in the fluid-solid interaction model. During simulations, the molten CMAS sub-domain flows through the Eulerian domain creating a moving free surface of the CMAS flow.

2.3.2. Boundary Conditions

The applied boundary conditions are summarized in Figure 7. The columns are kept undeformed by applying rigid body constraints. The symmetric boundary condition was applied at the sides, and the fixed boundary condition at the bottom (Figure 7a).

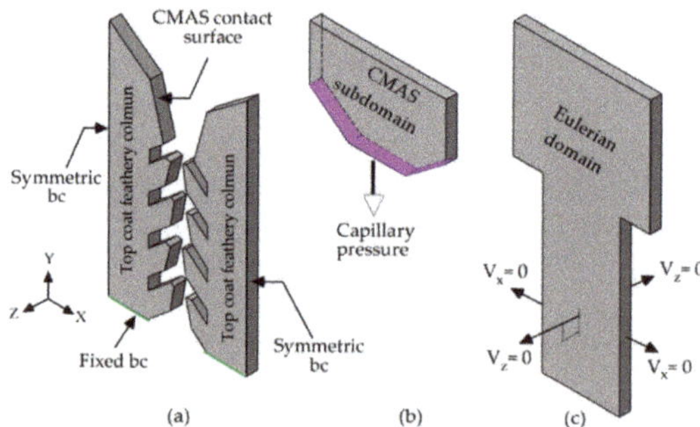

Figure 7. Boundary conditions for the fluid-solid interaction model (FSI): (**a**) boundary conditions for the rigid top coat model; (**b**) applied capillary pressure at the bottom surface of the molten CMAS; and (**c**) velocity boundary condition for the Eulerian domain.

To simulate the infiltration due to the capillary pressure along the feather gaps, a suction pressure (acting downwards) was imposed on the Eulerian subdomain (Figure 7b). An approximate value of the capillary pressure was calculated using Equation (1), as employed by [8] for an analytical infiltration analysis of the CMAS flow.

$$\Delta P = \frac{\sigma \cos \theta}{r_{\text{eff}}} \tag{1}$$

Here, θ is the wetting angle of the molten CMAS, σ is the surface tension of the liquid CMAS, and r_{eff} is the effective hollow radius between the columnar gaps. Assuming the fully wetted condition of CMAS, we obtained a zero wetting angle and $\cos\theta = 1$. The value of σ was taken as 0.438 N/m from the approximated value calculated for CMAS in [18], and r_{eff} was 18 μm from the microstructure analysis (Section 2.2.1).

Symmetry conditions of the Eulerian domain were implemented with velocity constraints in the required directions as follows: On X- and Z-planes the velocity is zero, i.e., $V_X = 0$ and $V_z = 0$. In the Y-direction the velocity was kept unconstraint allowing an equilibrium liquid flow through the top (inlet) and bottom (outlet) surfaces (Figure 7c). The contact surfaces between the columns and Eulerian domains are assumed to be slightly rough where the interactions between these surfaces were defined by a Coulomb friction model taking a small friction coefficient 0.1. This assumption was taken due to the lack of sufficient knowledge that characterizes the CMAS-TBC contact surface interactions during the CMAS flow. As the friction coefficient is small, the flow behavior will be mainly controlled by the viscosity parameter of the CMAS liquid. For all the models the friction coefficient was kept the same, therefore, the results from these models were comparable.

2.3.3. Constitutive Behavior and Material Parameters of CMAS

- Equation of State (EOS)

The properties of liquid silicate were described by an equation of state (EOS). With the EOS a quantitative functional relationship between the volumetric properties of a phase, and its specified composition, temperature (T) and pressure (P) can be established. In the present work, the EOS properties of the molten CMAS were approximated from the phase relationship of molten silicate: 'Diopside ($Ca_2Mg_2SiO_6$)–Anorthite ($CaAl_2Si_2O_8$) eutectoid liquid' due to the fundamental similarities of chemical compositions and physical behavior of CMAS [34,35].

For this particular molten CMAS, the complex relationships of their state variables, e.g., pressure, volume, and temperature were described by a Mie–Gruneisen type EOS material model as given in the ABAQUS software [33]. The constitutive model assumes the pressure as a function of the current density and internal energy per unit mass [$p = f(\rho, E_m)$], and takes the following form:

$$p = \frac{\rho c_0^2 \eta}{(1 - s\eta)^2}\left(1 - \frac{\Gamma_0 \eta}{2}\right) + \Gamma_0 \rho E_m \tag{2}$$

Here, p is the pressure, ρ is the reference density, η is the nominal volumetric compressive strain, E_m is the specific energy, and Γ_0 is the Grüneisen ratio. The parameter c_0 is the bulk speed of sound in the material and s is a parameter determined from the shock compression test by relating the shock velocity (U_s) and the particle velocity (U_p). Plotting the U_p in X-axis and U_s in Y-axis, a linear relationship of c_0 and s can be defined, which is known as the Hugoniot equation, and written as follows:

$$U_s = c_0 + sU_p \tag{3}$$

For the present work, the Hugoniot parameters for CMAS are approximated from the molten silicate data [36–38], which for the parameters c_0, s, and Γ_0 are: 3060 m/s, 1.36, and 1.52 respectively.

- Density and Viscosity of Molten CMAS

The density and viscosity of CMAS vary with their chemical compositions and are temperature dependent [39]. The density of laboratory produced CMAS in this work has been taken from a similar CMAS composition [40], where the density is reported to be in the range of 2.6–2.8 g/cc at 1300 °C. The approximated value for the present work is 2.69 g/cc at 1260 °C.

On the other hand, viscosity for the investigated CMAS was estimated by two methods: Experimental and theoretical. For a particular CMAS chemical composition (here, CMAS1: 41.68(SiO_2)–24.56(CaO)–12.43(MgO)–11.05(Al_2O_3)–8.71(FeO)–1.57(TiO_2)–0(SO_3) at.%) in the temperature range of 1200–1400 °C, the obtained viscosity shows a rapid decrease with increased temperature [26]. Due to the complicated data measurement techniques at high temperature, only one data was taken at each temperature. In a second approach, the viscosity was estimated theoretically using a well-established Giordano (GRD) empirical viscosity model that takes into account the compositional variation of CMAS and temperature dependency [41]. This model was verified on a large database of chemical compositions of molten silicates that were experimentally obtained.

In literature, the Giordano viscosity model has been widely used to estimate the temperature dependent viscosity of CMAS for a wide range of chemical compositions [42–45]. For the present work, the estimated GRD viscosity for the CMAS1 shows a lower viscosity compared to the experimental obtained viscosity data [31]. In Figure 8 both the experimentally estimated, and GRD model based viscosity is plotted for different temperature ranges. One can find that the viscosity prediction of the GRD model for CMAS1 at 1260 °C is ~0.5 Pa s, and the experimental value is close to 5.8 Pa s [26], which is almost 11 times higher.

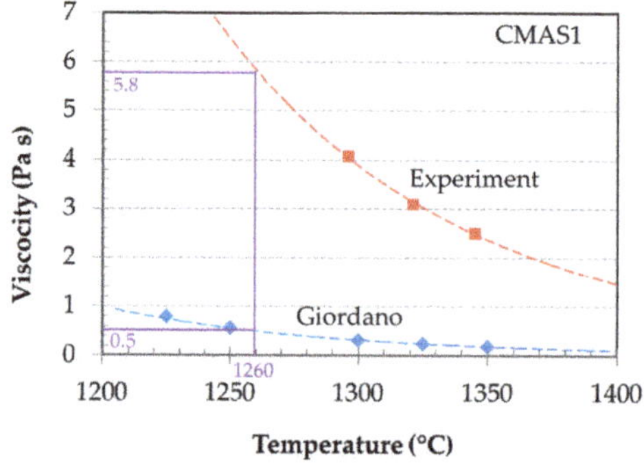

Figure 8. Viscosities for CMAS1 measured experimentally and using the Giordano (GRD) model at different temperatures.

For the numerical simulations, we have considered two arbitrary viscosity parameters, representative to the GRD (lower) and experimental values (higher), to obtain a qualitative understating of the influence of viscosities on the infiltration behavior. Both the parameters are shown in Figure 8 at the vertical axis. Several preliminary studies of the infiltration behavior were performed. Calculations with the lower viscosity parameter were numerically faster. Therefore, most of the analyses were performed with the low viscosity parameter, from which a fundamental comparative understanding of infiltration kinetics with respect to the EB-PVD morphology was obtained.

2.3.4. EB-PVD Microstructure Model for Infiltration Analysis

Geometric models having 40° and 60° feather inclinations with varying feather arm lengths (short and long), and columnar gaps (narrow and wide) were considered for the investigation of infiltration kinetics. For this purpose, six top coat model geometries were generated using a Python-based parametric model, where the morphological parameters were systematically varied. The models are shown in Figure 9. For each feather inclination angle (θ_F) two feather-arm gaps, G_F, and two columnar gaps, G_C were considered. Thus, changing the columnar gaps from narrow to wide and keeping the length of feather arms from short to long, the overall void vol.% (equivalent to porosity of the TBC) has been varied.

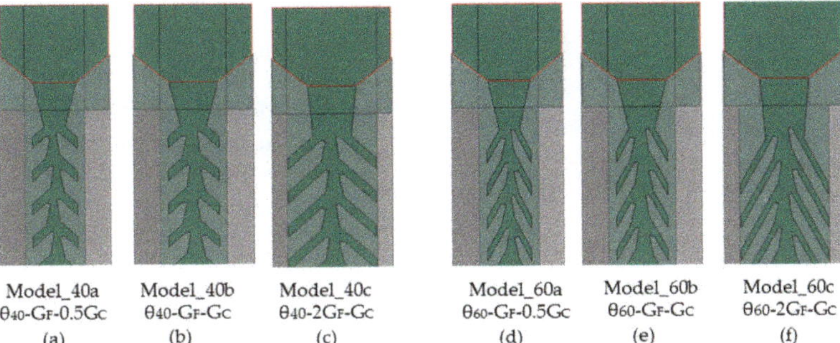

Model_40a   Model_40b   Model_40c   Model_60a   Model_60b   Model_60c

θ_{40}-G$_F$-0.5G$_C$ θ_{40}-G$_F$-G$_C$ θ_{40}-2G$_F$-G$_C$ θ_{60}-G$_F$-0.5G$_C$ θ_{60}-G$_F$-G$_C$ θ_{60}-2G$_F$-G$_C$

(a)   (b)   (c)   (d)   (e)   (f)

Figure 9. EB-PVD microstructure unit cell models including the TBC inter columnar gaps, molten CMAS domain and an Eulerian domain, generated using the Python script: (**a**) 40° inclined feather arm model with short arm and narrow columnar gap; (**b**) with short arm and wide columnar gap; (**c**) with long arm and wide columnar gap; (**d**) 60° inclined feather arm model with short arm and narrow columnar gap; (**e**) with short arm and wide columnar gap; and (**f**) with long arm and wide columnar gap.

The six models are named according to the variables used for this analysis: θ(Feather angle)–(Feather gap/void)G_F–(Columnar gap)G_C. Thus, for the 40° feather inclination model with double length feather arm and half columnar gap is denoted as: θ_{40}–2G_F–0.5G_C. Figure 9a–c shows the variants of 40° feather inclination models, and Figure 9c–e shows the variants of 60° feather inclination models.

It should be mentioned that the higher feather angle (60°) reduces the overall gaps between the feather arms, as can be seen in Figure 9b,e. During the model generation, the vertical gap parameter (V_F) was kept the same for 40° and 60° inclinations. However, due to the angular decomposition of V_F, the diameter of the feather gap for 60° inclined feathers (D_{60}) was reduced to $0.94D_{40}$, where D_{40} is the feather gap diameter at 40° inclination.

To study the CMAS flow behavior, data were evaluated along the columnar gap, represented with I1, I2, I3, and I4 sections, and inside the void between feather arms, represented with 1, 2, 3, and 4 numbers, in Figure 10.

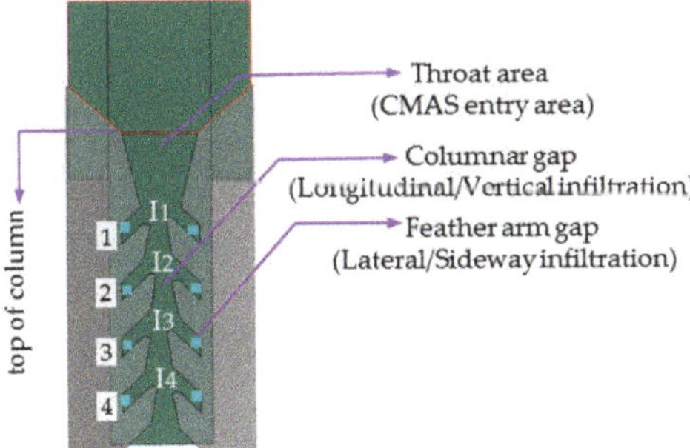

Figure 10. Defined sections at the columnar gap and void between feather arms for numerical data evaluation during infiltration kinetics study

The time occupied by the molten CMAS to fill one Eulerian element with EVF = 1 located at the flow front was taken to be the infiltration time for that element. The distance from this element to the top surface was taken as the depth (or height) of infiltration. The total infiltration time is the time required for filling up the empty Eulerian elements along any direction, which includes longitudinal (vertical) infiltration in the columnar gaps and lateral (sideway) infiltration in the feather arm voids.

103

It should be noted that the columnar height is the same for all the models but due to different feather morphology (e.g., feather inclination and length) the distance travelled by the molten CMAS within the feathery voids are not the same.

To obtain equivalent reliable numerical solutions from all the FE models, some pre-analyses were performed to optimize the mesh size, domain shape, and initial CMAS domain size. Only the optimized models were used for data evaluation and comparative analysis of the CMAS flow behavior. All simulations were performed using the low viscosity parameter until the CMAS infiltrated into the I4 section. For mutual comparison of infiltration depths, the data were evaluated after 80 µs of infiltration time.

In Figure 11, the infiltration along the longitudinal and lateral direction was shown for six models measured at 80 µs of infiltration time. Differences in the CMAS flow pattern can be well identified. The systematic analyses with the results are presented in the next section, where the infiltration kinetics with respect to the morphological variations is compared.

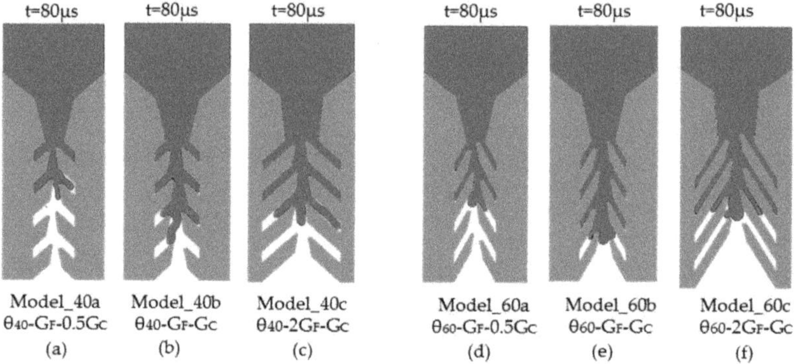

Figure 11. Infiltrated CMAS height at time of 80 µs simulated for six top coat morphologies. Infiltration depths are shown for: (**a**) Model_40a; (**b**) model_40b; (**c**) model_40c; (**d**) model_60a; (**e**) model_60b; and (**f**) model_60c.

3. Results

3.1. Morphological Influences on Infiltration Kinetics of CMAS

3.1.1. Influence of Feather Inclination and Feather Arm Length on CMAS Infiltration

Infiltration depths along the columnar gaps in the longitudinal (vertical) direction and along the void between feather arms (feather gaps) for lateral (sideway) infiltration were analyzed. The estimated total time for longitudinal infiltration involves the time required to fill up the vertical gaps in columns, preceded by filling up the sideway gaps between feathers. In Figure 12, the longitudinal infiltrations are summarized for four models, with feather arm angles at 40° and 60°, and with short and long feather gap. The columnar gap was the same for all models to gain a comparative overview of the morphological influences on the infiltration kinetics.

The columnar infiltration depth increased almost logarithmically with time, i.e., the infiltration was initially faster, and became slower as the CMAS approached towards the bottom sections. This effect came from the gradual increase of friction effects from the increased contact surfaces during material flow. Friction between the Eulerian material and column boundaries offers continuous resistance during contact, as a result the non-linear decrease of the flow was found. It can be seen that for the models with longer feather arms (or gaps), the overall vertical infiltration decreased and a lower infiltration depth was obtained for a particular infiltration time (for example, at 75 µs time). Comparing Figure 12a,b it can also be found that the infiltration depth is higher for 60° feather inclination compared to 40° inclined feathers. This result is compared with horizontal lines at an infiltration time of 75 μs in Figure 12a,b. The prediction indicates that a higher inclination angle of feather arms facilitates an overall easy flow of CMAS in the vertical direction.

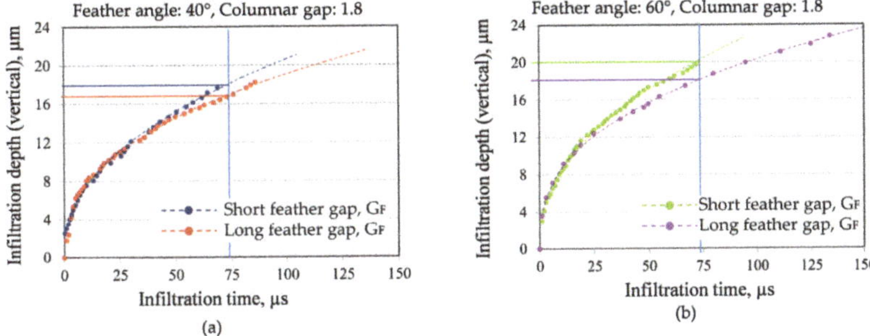

Figure 12. Vertical infiltration in columnar gaps with respect to geometrical factors variations, e.g., feather angle, feather arm length: (**a**) comparison for 40° inclined feathers with short and long arm; and (**b**) comparison for 60° inclined feathers with short and long arm.

The lateral (sideway) infiltration through long and short feather gaps was also analyzed in order to obtain a comparative view of local flow characteristics inside different types of feather gaps. The results are summarized in Figure 13 in terms of the feather position from the top (feather height) and time of infiltration. Data points were taken at the end of feather gaps in feather sets (FS), marked by 1, 2, 3, 4 in Figure 10. In these simulations, primary infiltration occurred vertically in the columnar gap, and in the course of the CMAS flow the secondary infiltration occurred sideways along the feather gaps. In Figure 13a,b, the results were compared for infiltration through feather gaps for the third feather-set FS3 for 40° and 60° feather angles. The apparent relationship between the infiltration time and infiltration depth inside the feather gaps are marked for both the cases. As shown in the plots, the CMAS flow inside the longer feather gaps requires higher time to fill-up. For the longer feather arm model, the dotted trend lines for data points (FS1–FS3) lies slightly above the trend line of the short feather arm model. The trend lines indicate that the infiltration depths at a particular time are slightly higher for longer feather gaps. This result is due to the slight differences in the contact surfaces for the long and short feather gaps. The short feather models exhibit slightly more tortuous flow path, which consists of several feather tips with complex geometry. The frictional effect on the flow front is affected by this geometry; as a result, slightly lower infiltration height was achieved in this model.

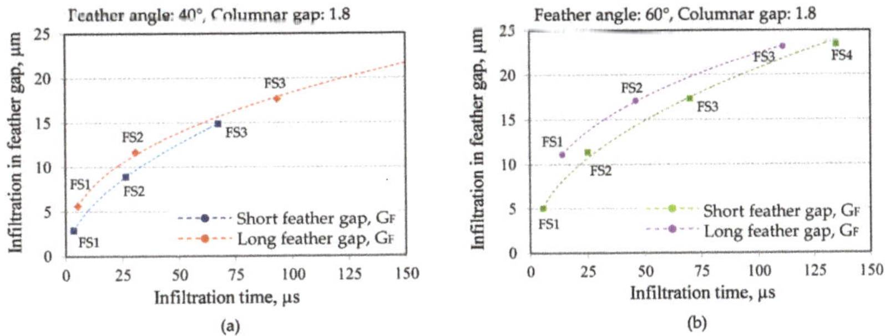

Figure 13. Lateral (sideway) infiltration in feather gaps with respect to geometrical factors variations, e.g., feather angle, feather arm length. Data are measured at four feather sets FS1–FS4: (**a**) comparison for 40° inclined feathers with short and long arm; (**b**) comparison for 60° inclined feathers with short and long arm.

3.1.2. Influence of Columnar Gap (Narrow and Wide) on CMAS Infiltration

The influence of the columnar gap on the infiltration behavior was studied for narrow ($G_C = 0.9$ μm) and wide ($G_C = 1.8$ μm) columnar gaps. Narrow columnar gap models were obtained simply by halving the inter-columnar gaps from the previous models. Models with short feather arms taking 40° and 60° feather inclinations were used. In Figure 14a,b, the infiltration depths with respect to time were summarized. For a particular infiltration depth large differences in the infiltration time were obtained between narrow and wide columnar gaps.

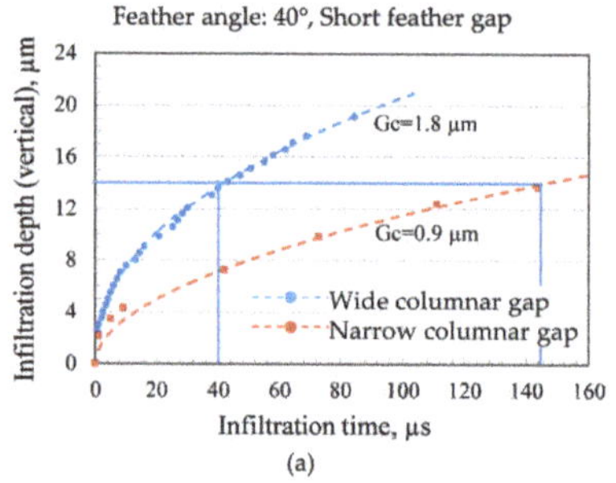

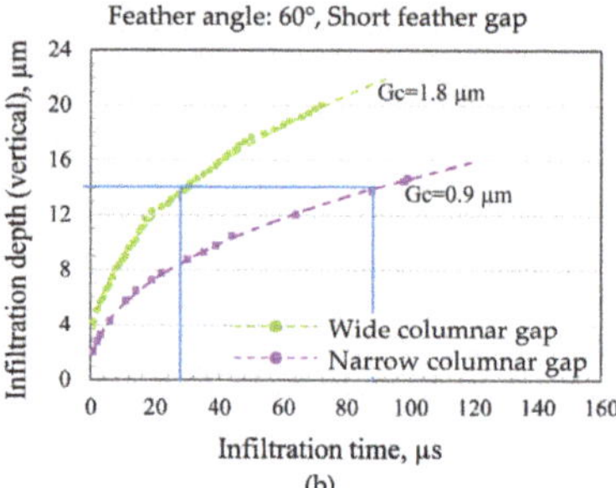

Figure 14. CMAS infiltration behavior for narrow and wide columnar gaps for short feather arm model: (**a**) Comparison for 40° inclined feather arm models; (**b**) comparison for 60° inclined feather arm models.

In Figure 14a, the infiltration time was almost 3.5 times higher for narrow columnar gaps to travel an infiltration depth of ca. 14 μm. By normalizing the data for total infiltration time, it can be shown that the infiltration time was increased by 250% in narrow columnar gaps ($G_C = 0.9$). The feather inclination was 40° for these models. Similarly, it can be shown from Figure 14b that the infiltration time was increased by ca. 200% in the narrow columnar gap for an infiltration depth of ca. 14 μm considering 60° feather inclination. The results showed that the reduction of the columnar gap width by a factor of 2.0 could reduce the infiltration time significantly. Comparing the results for feather inclinations one may also find that the models with 40° feather inclination showed reduced overall flow, which means, this model microstructure was more resistive to the molten CMAS infiltration compared to the 60° feather inclination model.

3.1.3. Microstructure Influence on the Infiltration Rate and Velocity

The morphological influence on the local infiltration kinetics can be understood clearly by analyzing the instantaneous rate of infiltration through columnar gaps for different feather morphologies. The instantaneous rate of infiltration was defined as the incremental change (Δh) in the columnar infiltration depth of the molten CMAS with respect to the incremental infiltration time (Δt). The infiltration rate was measured considering the morphological variations of feathery columns obtained by changing the feather arm size (for short or long $G_F = 1$ or 2) and feather inclination angle ($\theta_F = 40°$ or 60°).

In Figure 15, the infiltration rate with respect to the total time of infiltration was summarized for these models. From the plots (a–d) exponential decreasing trends of the curves were observed. The curves showed very high infiltration rate at the top of the columnar gaps just below the throat area. At this location only a little contact surface between TBC and CMAS was present, therefore, less frictional drag was activated allowing faster CMAS flow rate.

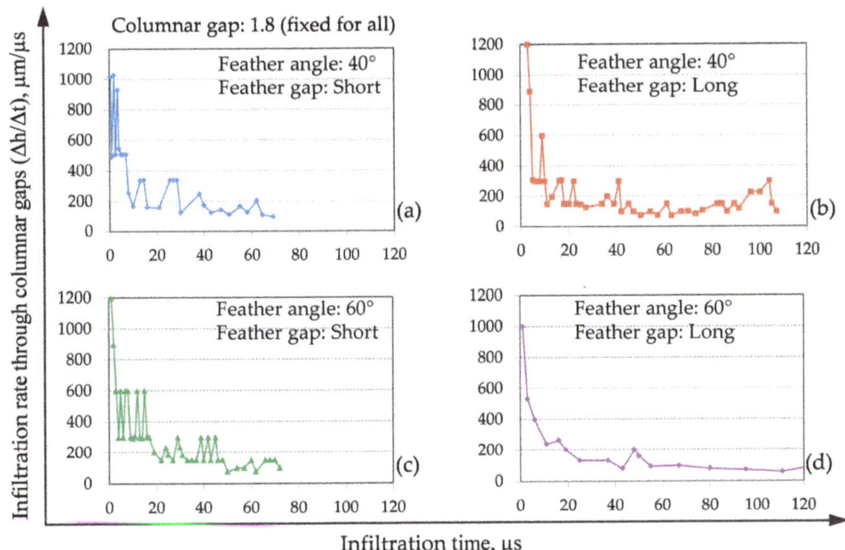

Figure 15. Change of infiltration rate through columnar gaps (vertical infiltration) influenced by different feather morphologies. The results are shown for: (**a**) short feathers with 40° feather inclination; (**b**) long feathers with 40° feather inclination; (**c**) short feathers with 60° feather inclination; and (**d**) long feathers with 60° feather inclination.

The rate of flow drops drastically when the contact surface was successively increased with the moving flow-front of CMAS. After filling up several feather gaps, the slope of the curves decreased gradually due to the decreasing flow rate and finally the rate became the slowest at the bottom of the columnar gap.

The rates of infiltration showed locally up and down zigzag patterns in the curves. These undulations were due to the interactions of CMAS at each feather-arm tips it encountered. At this location the primary flow path at the vertical direction put forth a new flow-path along lateral (sideway) direction through the feather gaps. Eventually, new contact surfaces came gradually in contact with the sideway flow which reduced the overall flow kinetics. As a resulting effect, the vertical infiltration became temporary stagnant or progressed with a minimal rate. Once the feather gaps were filled with CMAS, the vertical flow became faster again. The zigzag pattern was the result of slow-fast infiltration rates along the columnar gaps due to the branching of the CMAS flow path. The slow rate was induced by the lateral movement of CMAS through feather gaps having different morphological variations.

For a comparative study of the overall rate of the vertical infiltration with respect to the variations of feather arm length and inclination, the previous flow curves were reorganized with respect to the trend-line behavior of infiltration kinetics. The trend-lines basically approximate the curve behavior by fitting polynomial functions based on the available data points, thus the zigzag data patterns were idealized. In Figure 16, the morphological influences on the flow behavior can be recognized more clearly, from which two conclusive results can be drawn: (i) changing the feather inclination angle from 40° to 60°, the vertical infiltration rate (overall infiltration) at column gaps became faster; and (ii) increasing the feather length from short to long the vertical infiltration rate becomes slower.

For the first case, the 60° feather inclination narrowed down the feather gap diameter slightly (can be seen in Figure 9). As a consequence, the frictional drag per unit area was increased; therefore, the CMAS flow rate was reduced inside the feather gaps. However, the overall porosity is reduced, therefore, during less lateral flow in the feather gaps, more liquid flows into the vertical gaps, resulting in a faster flow rate (on average) compared to the 40° feather angle model.

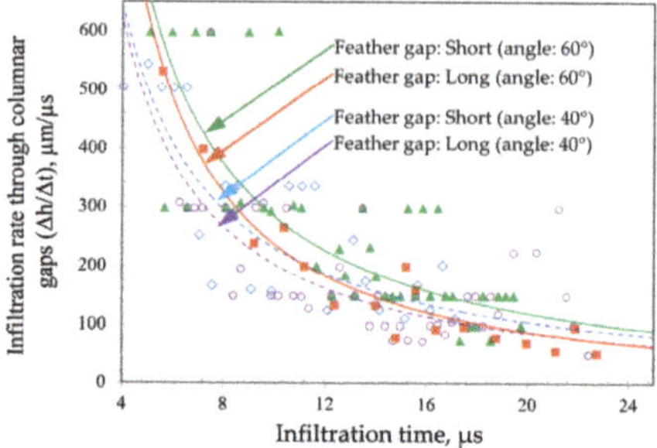

Figure 16. Simulated infiltration rates in columnar gaps compared for short and long feather arm models considering 40° and 60° feather arm inclination.

For the second case, the longer feather arm increased the feather gaps. The frictional contact surface that the flow-path encounters becomes longer as well. In this model, during the CMAS flow the vertical flow-path subdivided into wider and longer branches along the feather gaps, reducing the effective flow rate along the vertical direction.

To understand the overall rate of lateral (sideway) infiltration, the infiltration velocity was computed along the feather gaps for each feather set (FS). An average infiltration velocity was estimated by measuring the required time for distance travelled by CMAS at the feather gaps. In Figure 17, the results were summarized for the four models. As the distances of the feather arm sets (FS) from the top position were different, different infiltration depth for FS1–FS4 was obtained. To compare the results, the trend-line behavior of the velocity data was analyzed.

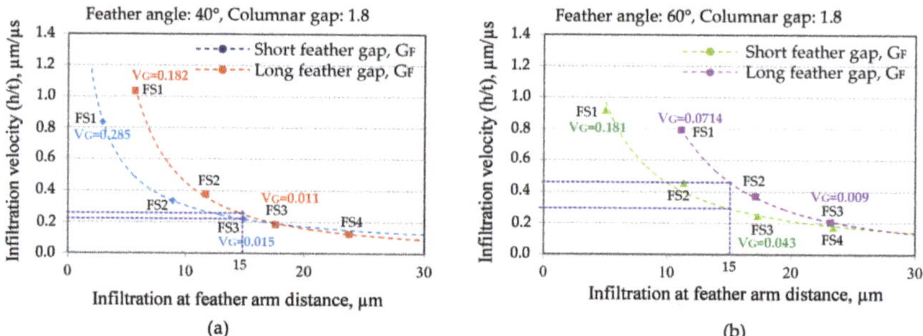

Figure 17. Lateral (sideway) infiltration in feather gaps showing velocity reduction in feather arm sets (FS) and the change of velocity gradient: (**a**) Comparison for 40° inclined feathers; (**b**) comparison for 60° inclined feathers.

The trend-lines showed that the flow was faster at the first feather gaps, and with the course of time, it was reduced successively for the second and third feather gaps. The reduction rate of the velocity depends on the frictional surface areas against which the CMAS interacts during flowing. For longer feather gaps the velocity reduction is higher, as the CMAS has to travel longer distances by overcoming longer frictional flow paths.

It should be noted that the high velocity (faster flow) at the FS1 is influenced by the wide CMAS entry area, which is slightly different for each model (recall Figure 11). The feather set FS3 is away from the CMAS entry area, therefore, at this position the flow becomes steady. This steadiness with respect to the feather gap size and inclination is compared with the velocity gradient ($V_G = \Delta v/\Delta h$) data, as shown in Figure 17.

3.2. Influence of Viscosity on CMAS Infiltration

The influence of viscosity on the flow behavior for long feather arms and 40° feather inclination (model: θ_{40}–2G_F–G_C) was investigated by using two viscosity values: Low and high as stated in Section 2.3.3. The characteristics of the flow kinetics were distinguished by comparing the flow rate compared to the infiltration depth relationship, as shown in Figure 18. Here, the instantaneous rate of infiltration ($\Delta h/\Delta t$) was plotted, which captured the local change of the infiltration rate during traveling through the columnar gaps. The instantaneous rate gives a better understanding of the flow kinetics than the average rate of change, as for the earlier case the local variation can be better explored and understood.

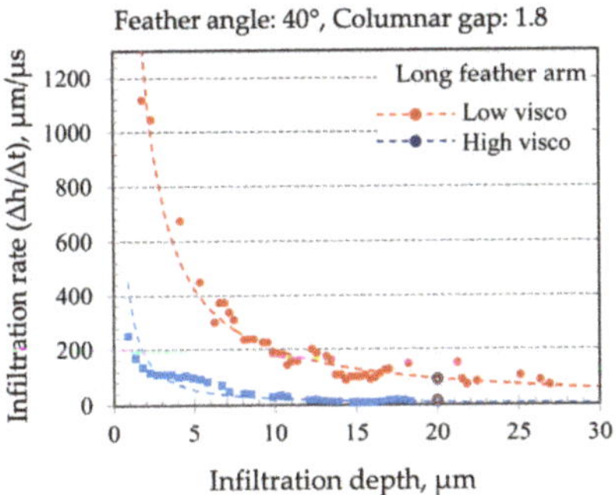

Figure 18. Decrease in the infiltration rate with respect to the infiltration depth measured in columnar gaps based on low and high viscosity values for 40° feather inclination, 1.8 µm columnar gap and long feather arm condition.

As seen in Figure 18, the low viscous CMAS entered the column with a high flow rate and then the flow was gradually subsided while filling up the columnar gap. On the other hand, the high viscous fluid entered the columnar gap with an almost 8–9 times slower flow rate, as shown at a distance of 2 µm infiltration depth in the figure. After filling the columnar and feathery gaps at a distance of 20 µm, the higher viscos CMAS almost became sluggish, but the low viscous CMAS continued further with a slower flow rate.

For the high viscosity parameter a higher flow resistance was imposed on the molten CMAS, which eventually reduces the flow velocity inside the columnar and feathery gaps.

A comparison of the flow velocity for low and high viscosity parameters along the columnar gaps shows a similar decreasing pattern as given in Figure 18, and therefore, not presented here.

Further, the lateral (sideway) infiltration behavior for low and high viscosity parameters was studied for two different morphologies: One is for short and the other is for long feather arms (model. θ_{40}–G_F–G_C and θ_{40}–2G_F–G_C), with the feather inclination angle at 40°. The infiltration time required for filling up the void between feather arms considering short and long feather arms were compared. In Figure 19, the total infiltration time was recorded for filling up each feather set (FS) in the θ_{40}–G_F–G_C and θ_{40}–2G_F–G_C model. This infiltration time also includes the time of vertical infiltration, as the vertical and horizontal infiltration occurs successively.

In Figure 19a, the short feather arm model shows a slower infiltration for a high viscous CMAS compared to a lower viscous CMAS, as expected. The dotted lines indicate that the infiltration time increased exponentially for the high viscous CMAS while filling the longitudinal and lateral voids (in columnar and feathery gaps), while a moderate increase of time was observed for the low viscous CMAS. For the low viscous CMAS (Figure 19b), the infiltration time remained qualitatively the same for both short and long feather gaps. For the high viscous CMAS, the infiltration time increased remarkably for long feather gaps compared to the shorter one.

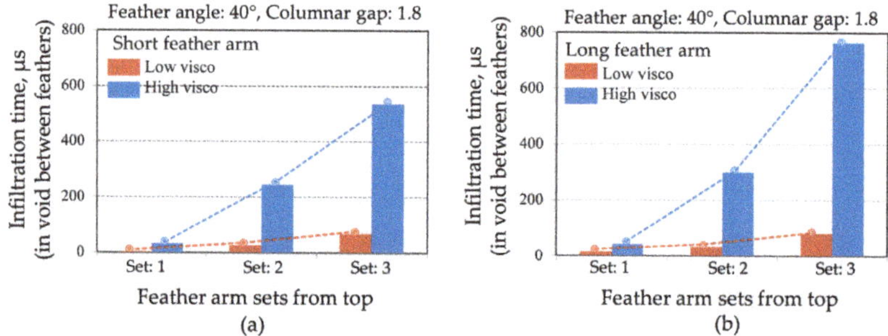

Figure 19. Comparison of CMAS infiltration into the feather gaps for different feather sets using a low and a high viscosity parameter simulated in a model with 40° feather inclination and 1.8 μm columnar gap: (**a**) comparison of infiltration time to fill up three short feather gaps; and (**b**) comparison of infiltration time to fill up three long feather gaps.

3.3. Qualitative Comparison with Experimental Infiltration Behavior

The CMAS infiltration kinetics was studied experimentally in previous researches [19,24] for the case of coarse and feathery morphologies (see Figure 2b,c). The overall porosity (gaps between columns and feathers) in the feathery structure was almost 2.5 times higher than that of the coarse structure. The main goal of the experimental investigation was to identify which microstructure resists CMAS infiltration depending on the geometrical factors. During the experiment infiltration was carried out for different times at a constant temperature. Then for both microstructures, the infiltration depth was measured by means of SEM imaging with the help of the energy-dispersive X-ray spectroscopy (EDS) analysis. The maximum infiltration depth of CMAS is normalized for both microstructures for a comparative study. If the maximum infiltration depth of CMAS is considered to be 100% (for a coating thickness of 400 μm), and the maximum infiltration time is 100%, then the infiltration rates for coarse and feathery structures can be qualitatively compared, which are shown in Figure 20a.

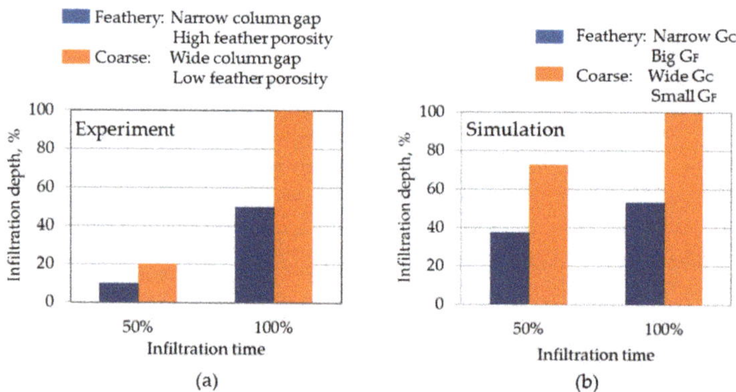

Figure 20. Comparison of infiltration depths and infiltration times for 'feathery' and 'coarse' morphology: (**a**) experimental results (normalized) at two columnar distances; and (**b**) simulation based prediction (normalized) at two columnar distances.

It was found that after 50% and 100% infiltration time, the difference in infiltration depth between feathery to coarse microstructure was found to be 50%. At any point of the infiltration time the feathery morphology exhibits almost 50% lesser infiltration depth.

For the numerical study, an equivalent microstructure for coarse and feathery columns was obtained from the θ_{40}–G_F–G_C and θ_{40}–$2G_F$–$0.5G_C$ model. For both the models, simulations were performed only up to 15 µm infiltration depths, which cover full infiltration beyond four feather-sets. The respective variations of infiltration depths for two intervals of the infiltration time similar to the experimental case were compared. The results are depicted in Figure 20b. The difference of infiltration depth was found to be 46.8% at 100% infiltration time, which is close to the experimental result. At 50% of the maximum infiltration time the difference was found to be 48.21%. In both experimental and simulation results the infiltration depth was found to be ~50% lower in the feathery (feathers with double length) microstructure.

The numerical prediction of infiltration depth at 100% infiltration time predicts the experimental observation reasonably well, but for 50% infiltration time the model showed higher infiltration depth compared to experiments. This discrepancy can be explained on the basis of the melting nature of the CMAS melt. It is to be noted that at 1225 °C, the CMAS melt was in semi molten state (crystalline + glass) and longer heating times were required to make it fully glassy in nature which was the case of 20 h infiltration results (100% infiltration at 1225 °C). For the model simplicity, it was assumed that the whole CMAS was in a glassy and flowing condition, which in turn has resulted in higher infiltration depth than in the experiments. However, it was shown that the comparative trend of infiltration kinetics matches very well with the experimental observation.

4. Discussion

The CMAS infiltration kinetics was assessed for an EB-PVD TBC microstructure with varying morphology, which includes geometrical features for feather arm length, feather inclination angle, feather gaps, columnar gaps, etc. Based on the model parameters the infiltration behavior was quantitatively estimated. Some qualitative understanding of the infiltration kinetics with respect to the morphological variations was also gained. Some important results are discussed as following.

4.1. Influence of Columnar Gap width on Infiltration Behavior

Columnar gaps were arranged vertically in the models as described previously in Figures 3 and 4. The tortuosity along the column height was modeled by arranging the longer and shorter feather arms alternatively. The negative pressure force (equivalent to a constant capillary force) was active in this gap. The infiltration flow was obstructed by the long and short feather arms with sharp feathery tips, and branched by the feather gaps with different inclination. The contact surfaces imposed frictional drag during flow progression. For a wide columnar gap the unit frictional surface area with respect to the total surface area is reduced because the ratio of the friction surface length and the columnar diameter (L/D_C) is less compared to a narrow columnar gap. That means, for wide columnar gaps a unit volume of liquid is exposed to less frictional drag during infiltrating. For a narrow columnar gap the overall porosity is reduced, and the frictional effect is significantly higher for a unit volume of liquid to enter, which obviously increases the infiltration time (Figure 14).

4.2. Influence of Feather Length (Gap) on Total Infiltration Behavior

The vertical infiltration time estimated in the models is the time required to fill-up all the voids, e.g., columnar gaps arranged in vertical directions and feather gaps arranged laterally (sideway). Therefore, the vertical infiltration time is considered as the time of total infiltration. It was shown that for a constant columnar gap the variations in feather morphology, e.g., feather arm lengths and feather inclination influence the infiltration kinetics remarkably.

The feather arm length basically defines the length of feather gaps between two feathers. For a set of long feather arms the void between feather arms increases. Infiltration time is expected to be higher for this type of long feather model. Similar behavior was also observed in the experiments comparing coarse and feathery structures of the top coat columns [19,24]

As the contact surface areas between the CMAS flow path and the top coat boundaries retard the flow rate due to frictional effects, for an increased contact surface area slow infiltration was obtained.

Therefore, a large feather arm model containing large feather gaps and large contact surface areas showed reduced overall flow rate.

Another morphological factor, the feather inclination, also affects the infiltration kinetics, basically due to the narrow feather gaps obtained during model generation. Narrow feather gaps increase the frictional drag effects per unit area as discussed in Section 3.1.3. As a result, the CMAS flow rate inside a more inclined feather gap was reduced. For 40° inclined feathers the gap was wider, and thus larger, therefore, the acted frictional drag on CMAS was less, and inside the feather gap flow was faster.

However, for the 60° inclined feather model the overall porosity is smaller compared to the 40° inclined feather as the both have the same inter-columnar gaps. As a result, the vertical flow rate, which is the resultant of the flow along inter-columnar and feather gaps, was higher for the 60° inclined model.

It should be noted that the narrow feather gap for a higher feather angle is only due to the trigonometric conversation of data in the parametric model output. In reality, an inclined feather can possess any arbitrary feather gaps, not necessarily a narrower feather gap.

4.3. Influence of Morphology on the Flow Velocity and Rate

For all of the models the rate of infiltration was higher at the very beginning, and it was reduced gradually with respect to the vertical infiltration along the columnar gaps. This behavior was obvious as the CMAS entry area offered less frictional contacts, and for the progressive CMAS flow increased frictional contact areas were activated on the fly.

It was shown that the flow velocity in feather gaps and columnar gaps are dependent on the gap diameter, which determines the vol.% of porosity and effective frictional surface encountered by the flow-front. For both the columnar and feather gaps the flow velocity was reduced for narrower gaps due to large frictional effects per unit area. For a more vertically inclined feather arms (e.g., 60° inclined), the flow at the narrow feather gaps was reduced but the vertical infiltration occurred continuously showing an overall increased flow rate (Section 3.1.3). Therefore, we observed higher infiltration depth in the 60° feather angle model at a particular infiltration time (Figure 12).

4.4. Influence of Viscosity

The known effects of viscosity on the CMAS flow kinetics in TBCs were once again proven with the developed model showing that the CMAS with less viscosity has the faster infiltration rates. For a high viscous flow, the effects of frictional drag increase exponentially with the distance travelled by the CMAS flow. For the low viscosity parameter, the frictional drag increases only slightly during continuous flow along the frictional flow path.

No linear relationship between the flow rates can be found comparing the results of low and high viscosity values. That means, a linear increase of viscosity parameter will not increase the infiltration time linearly; rather an exponential increase of infiltration time can be expected.

4.5. Suggestions and Remarks toward Future Direction

The presented simulations captured a quantitative picture of microstructure sensitivity of the CMAS infiltration behavior. Valuable information can be extracted from the predictive analysis, from which the critical morphological features that need to be customized for mitigating CMAS infiltration can be identified.

For example, the results obtained from the six idealized 2D models suggest that EB-PVD top coats may comprise some morphological features, like narrow inter-columnar gaps, long feather arms, lower feather inclination angles, and above all higher lateral porosity, in order to reduce the CMAS intake through infiltration. Simulation-based knowledge can thus be used for advanced TBC design against environmental attack.

It should be noted that the predictability of the present model is restricted due to several idealizations in flow behavior and morphology. These limitations can be improved by employing detailed physics-based modeling incorporating complex 3D representations of the TBC microstructures. In particular, a thermodynamic equilibrium based melt/coating interaction model taking the thermal gradient as well as the temperature dependent viscous flow of CMAS, considering crystallization dynamics during infiltration needs to be developed. Realization of such thermo-chemical interaction modeling requires an enhancement of the presented multi-physical computational approach, which is currently in progress and remains beyond the scope of this work.

The computational method presented here is not limited to the CMAS-TBC interaction analysis only. This modeling approach can be extended for the analysis of CMAS-EBC (environmental barrier coating) interactions with few modifications in the microstructure model.

5. Conclusions

In this work, a numerical approach was developed for studying the infiltration kinetics of molten CMAS considering the morphological variation of EB-PVD TBCs. For a numerical representation of the top coat TBC microstructure, a microstructure model was established by rational idealization of morphological features from the SEM images. Six models were generated for the TBC-CMAS interaction analysis. Detailed analyses of the flow kinetics were performed, and then the results were verified by experimental infiltration depths of the feathery and coarse microstructures. Based on the predicted results the flow characteristics regarding infiltration depth, rate, and velocity were evaluated with respect to the columnar microstructural features of the top coat, enabling a qualitative comparison of the microstructure sensitive flow kinetics of CMAS. An in-depth understanding of the CMAS flow behavior was gained, and several morphological features of the EB-PVD TBC controlling infiltration kinetics were identified. The results are summarized as follows:

- Size of feather arms: For long feather arms the void between feather arms is elongated, thereby pore vol.% is increased. Elongated voids consist of large frictional surfaces that induce a higher drag against the CMAS flow. As a result, for the long feather arms the overall flow rate of CMAS can be reduced.
- Feather arm inclination: For a higher inclination of feather arms the gap between feather arms becomes narrower due to the angular projection of the feather gap diameter. A narrow gap obviously imposes more frictional drag per unit area; therefore, the flow rate of CMAS in the feather gaps (lateral infiltration) is reduced. Keeping the vertical columnar gaps the same, the overall porosity decreases in a narrow feather gap model and oppositely, increases in a wide feather gap model. It was found that the flow rate at the vertical columnar gaps (longitudinal infiltration) remains high due to the reduction of overall porosity in the feather gaps when compared to the wider feather gap model.
- Columnar gaps: Keeping the feather lengths the same, in a model with narrow columnar gaps the overall porosity decreases and the frictional drag per unit area increases. Thus, reducing the diameter between columnar channels a retardation of CMAS flow can be expected. If the columnar diameter is the same for two models, the model with a lower pore volume percent will take less infiltration time.

In future, the microstructure sensitive flow kinetics of molten CMAS will be assessed using 3D microstructure models considering extended geometrical features of the TBC columns. Furthermore, the model can be used for the thermomechanical stress analysis to predict the CMAS assisted stress evolution and fracture in advanced TBC layers.

Author Contributions: Conceptualization, M.R.K. and R.N.; Methodology, M.R.K.; Validation, M.R.K. and N.R.; Formal Analysis, A.K.S.; Investigation, M.R.K., A.K.S. and R.N.; Writing–Original Draft Preparation, M.R.K.; Writing–Review and Editing, A.K.S., R.N. and U.S.; Supervision, R.N. and U.S.; Project Administration, R.N. and U.S.; Funding Acquisition, R.N. and U.S.

Funding: This research was funded by Deutsche Forschungsgemeinschaft (DFG) (No. Schu 1372/5-1).

Acknowledgments: The authors acknowledge the supports from ICAMS, Ruhr-University Bochum for providing short-term computer facilities.

Conflicts of Interest: The authors declare no conflict of interest. The funders had no role in the design of the study; in the collection, analyses, or interpretation of data; in the writing of the manuscript, or in the decision to publish the results.

References

1. Leyens, C.; Schulz, U.; Pint, B.A.; Wright, I.G. Influence of EB-PVD TBC microstructure on thermal barrier coating system performance under cyclic oxidation conditions. *Surf. Coat. Technol.* **1999**, *120*, 68–76. [CrossRef]
2. Evans, A.G.; Mumm, D.R.; Hutchinson, J.W.; Meier, G.H.; Pettit, F.S. Mechanisms controlling the durability of thermal barrier coatings. *Prog. Mater. Sci.* **2001**, *46*, 505–553. [CrossRef]
3. Schulz, U.; Leyens, C.; Fritscher, K.; Peters, M.; Saruhan-Brings, B.; Lavigne, O.; Dorvaux, J.-M.; Poulain, M.; Mévrel, R.; Caliez, M. Some recent trends in research and technology of advanced thermal barrier coatings. *Aerosp. Sci. Technol.* **2003**, *7*, 73–80. [CrossRef]
4. Clarke, D.; Levi, C. Materials design for the next generation thermal barrier coatings. *Annu. Rev. Mater. Res.* **2003**, *33*, 383–417. [CrossRef]
5. Levi, C.G. Emerging materials and processes for thermal barrier systems. *Curr. Opin. Solid State Mater. Sci.* **2004**, *8*, 77–91. [CrossRef]
6. Clarke, D.R.; Oechsner, M.; Padture, N.P. Thermal-barrier coatings for more efficient gas-turbine engines. *MRS Bull.* **2012**, *37*, 891–898. [CrossRef]
7. Krämer, S.; Yang, J.; Levi, C.G.; Johnson, C.A. Thermochemical interaction of thermal barrier coatings with molten CaO–MgO–Al$_2$O$_3$–SiO$_2$ (CMAS) deposits. *J. Am. Ceram. Soc.* **2006**, *89*, 3167–3175. [CrossRef]
8. Zhao, H.; Levi, C.G.; Wadley, H.N.G. Molten silicate interactions with thermal barrier coatings. *Surf. Coat. Technol.* **2014**, *251*, 74–86. [CrossRef]
9. Borom, M.P.; Johnson, C.A.; Peluso, L.A. Role of environment deposits and operating surface temperature in spallation of air plasma sprayed thermal barrier coatings. *Surf. Coat. Technol.* **1996**, *86–87*, 116–126. [CrossRef]
10. Mercer, C.; Faulhaber, S.; Evans, A.G.; Darolia, R. A delamination mechanism for thermal barrier coatings subject to calcium–magnesium–alumino–silicate (CMAS) infiltration. *Acta Mater.* **2005**, *53*, 1029–1039. [CrossRef]
11. Grant, K.M.; Krämer, S.; Löfvander, J.P.A.; Levi, C.G. CMAS degradation of environmental barrier coatings. *Surf. Coat. Technol.* **2007**, *202*, 653–657. [CrossRef]
12. Wellman, R.; Whitman, G.; Nicholls, J.R. CMAS corrosion of EB PVD TBCs: Identifying the minimum level to initiate damage. *Int. J. Refract. Met. Hard Mater.* **2010**, *28*, 124–132. [CrossRef]
13. Levi, C.G.; Hutchinson, J.W.; Vidal-Sétif, M.-H.; Johnson, C.A. Environmental degradation of thermal-barrier coatings by molten deposits. *MRS Bull.* **2012**, *37*, 932–941. [CrossRef]
14. Aygun, A.; Vasiliev, A.L.; Padture, N.P.; Ma, X. Novel thermal barrier coatings that are resistant to high-temperature attack by glassy deposits. *Acta Mater.* **2007**, *55*, 6734–6745. [CrossRef]
15. Krämer, S.; Yang, J.; Levi, C.G. Infiltration-inhibiting reaction of gadolinium zirconate thermal barrier coatings with CMAS melts. *J. Am. Ceram. Soc.* **2008**, *91*, 576–583. [CrossRef]
16. Rai, A.K.; Bhattacharya, R.S.; Wolfe, D.E.; Eden, T.J. CMAS-resistant thermal barrier coatings (TBC). *Int. J. Appl. Ceram. Technol.* **2010**, *7*, 662–674. [CrossRef]
17. Pubbysetty, R.P. Developing the CMAS (CaO–MgO–Al$_2$O$_3$–SiO$_2$) Resistant TBCs (Thermal Barrier Coatings): Alumina (Al$_2$O$_3$) as A Candidate for Restricting the CMAS Infiltration in TBCs. Master's Thesis, RWTH Aachen, Aachen, Germany, July 2016.
18. Gomez Chavez, J.J. Yttria Rich TBCs as Candidates for CMAS Resistant Top Coats. Master's Thesis, University of Texas, El Paso, TX, USA, May 2016.
19. Naraparaju, R.; Gomez Chavez, J.J.; Schulz, U.; Ramana, C.V. Interaction and infiltration behavior of Eyjafjallajökull, Sakurajima volcanic ashes and a synthetic CMAS containing FeO with/in EB-PVD ZrO$_2$–65 wt % Y$_2$O$_3$ coating at high temperature. *Acta Mater.* **2017**, *136*, 164–180. [CrossRef]

20. Drexler, J.M.; Gledhill, A.D.; Shinoda, K.; Vasiliev, A.L.; Reddy, K.M.; Sampath, S.; Padture, N.P. Jet engine coatings for resisting volcanic ash damage. *Adv. Mater.* **2011**, *23*, 2419–2424. [CrossRef] [PubMed]

21. Schulz, U.; Braue, W. Degradation of $La_2Zr_2O_7$ and other novel EB-PVD thermal barrier coatings by CMAS (CaO–MgO–Al_2O_3–SiO_2) and volcanic ash deposits. *Surf. Coat. Technol.* **2013**, *235*, 165–173. [CrossRef]

22. Mechnich, P.; Braue, W. Volcanic ash-induced decomposition of EB-PVD $Gd_2Zr_2O_7$ thermal barrier coatings to Gd-oxyapatite, zircon, and Gd, Fe–zirconolite. *J. Am. Ceram. Soc.* **2013**, *96*, 1958–1965. [CrossRef]

23. Eils, N.K.; Mechnich, P.; Braue, W. Effect of CMAS deposits on MOCVD coatings in the system Y_2O_3–ZrO_2: Phase relationships. *J. Am. Ceram. Soc.* **2013**, *96*, 3333–3340. [CrossRef]

24. Naraparaju, R.; Hüttermann, M.; Schulz, U.; Mechnich, P. Tailoring the EB-PVD columnar microstructure to mitigate the infiltration of CMAS in 7YSZ thermal barrier coatings. *J. Eur. Ceram. Soc.* **2017**, *37*, 261–270. [CrossRef]

25. Renteria, A.F.; Saruhan, B.; Schulz, U.; Raetzer-Scheibe, H.-J.; Haug, J.; Wiedenmann, A. Effect of morphology on thermal conductivity of EB-PVD PYSZ TBCs. *Surf. Coat. Technol.* **2006**, *201*, 2611–2620. [CrossRef]

26. Naraparaju, R.; Pubbysetty, R.P.; Mechnich, P.; Schulz, U. EB-PVD alumina (Al_2O_3) as a top coat on 7YSZ TBCs against CMAS/VA infiltration: Deposition and reaction mechanisms. *J. Eur. Ceram. Soc.* **2018**, *38*, 3333–3346. [CrossRef]

27. Caliez, M.; Feyel, F.; Kruch, S.; Chaboche, J.-L. Oxidation induced stress fields in an EB-PVD thermal barrier coating. *Surf. Coat. Technol.* **2002**, *157*, 103–110. [CrossRef]

28. Bhatnagar, H.; Ghosh, S.; Walter, M.E. A parametric study of damage initiation and propagation in EB-PVD thermal barrier coatings. *Mech. Mater.* **2010**, *42*, 96–107. [CrossRef]

29. Zhang, X.; Xu, B.; Wang, H.; Wu, Y. An analytical model for predicting thermal residual stresses in multilayer coating systems. *Thin Solid Films* **2005**, *488*, 274–282. [CrossRef]

30. Hochstein, L. Lebensdauer von 7YSZ EB-PVD Wärmedämmschichten in Flugtriebwerken mit Lokaler CMAS Belastung: Ausfallmechanismen und Modellierung. Master's Thesis, TU Dresden, Dresden, Germany, March 2017.

31. Naraparaju, R.; Gomez Chavez, J.J.; Niemeyer, P.; Hess, K.-U.; Song, W.; Dingwell, D.B.; Lokachari, S.; Ramana, C.V.; Schulz, U. Estimation of CMAS Infiltration depth in EB-PVD TBCs: A new constraint model supported with experimental approach. *J. Eur. Ceram. Soc.* **2019**, *39*, 2936–3945. [CrossRef]

32. Schulz, U.; Terry, S.G.; Levi, C.G. Microstructure and texture of EB-PVD TBCs grown under different rotation modes. *Mater. Sci. Eng. A* **2003**, *360*, 319–329. [CrossRef]

33. *ABAQUS*, V., 6.16; Documentation, Theory manual; Dassault Systèmes Simulia Corp.: Providence, RI, USA, 2016.

34. Barbieri, L.; Bondioli, F.; Lancellotti, I.; Leonelli, C.; Montorsi, M.; Ferrari, A.M.; Miselli, P. The anorthite–diopside system: Structural and devitrification study. Part II: Crystallinity analysis by the Rietveld–RIR method. *J. Am. Ceram. Soc.* **2005**, *88*, 3131–3136. [CrossRef]

35. Getson, J.M.; Whittington, A.G. Liquid and magma viscosity in the anorthite-forsterite-diopside-quartz system and implications for the viscosity-temperature paths of cooling magmas. *J. Geophys. Res. Solid Earth* **2007**, *112*, B10203. [CrossRef]

36. Rigden, S.M.; Ahrens, T.J.; Stolper, E.M. Shock compression of molten silicate: Results for a model basaltic composition. *J. Geophys. Res. Solid Earth* **1988**, *93*, 367–382. [CrossRef]

37. Ai, Y.; Lange, R.A. New acoustic velocity measurements on CaO–MgO–Al_2O_3–SiO_2 liquids: Reevaluation of the volume and compressibility of $CaMgSi_2O_6$–$CaAl_2Si_2O_8$ liquids to 25 GPa. *J. Geophys. Res. Solid Earth* **2008**, *113*, B04203. [CrossRef]

38. Thomas, C.W.; Asimow, P.D. Direct shock compression experiments on premolten forsterite and progress toward a consistent high-pressure equation of state for CaO–MgO–Al_2O_3–SiO_2–FeO liquids. *J. Geophys. Res. Solid Earth* **2013**, *118*, 5738–5752. [CrossRef]

39. Wiesner, V.L.; Bansal, N.P. Mechanical and thermal properties of calcium–magnesium aluminosilicate (CMAS) glass. *J. Eur. Ceram. Soc.* **2015**, *35*, 2907–2914. [CrossRef]

40. Sparks, R.S.J.; Huppert, H.E. Density changes during the fractional crystallization of basaltic magmas: Fluid dynamic implications. *Contrib. Miner. Pet.* **1984**, *85*, 300–309. [CrossRef]

41. Giordano, D.; Russell, J.K.; Dingwell, D.B. Viscosity of magmatic liquids: A model. *Earth Planet. Sci. Lett.* **2008**, *271*, 123–134. [CrossRef]

42. Wiesner, V.L.; Vempati, U.K.; Bansal, N.P. High temperature viscosity of calcium–magnesium–aluminosilicate glass from synthetic sand. *Scr. Mater.* **2016**, *124*, 189–192. [CrossRef]

43. Kumar, R.; Jordan, E.; Gell, M.; Roth, J.; Jiang, C.; Wang, J.; Rommel, S. CMAS behavior of yttrium aluminum garnet (YAG) and yttria-stabilized zirconia (YSZ) thermal barrier coatings. *Surf. Coat. Technol.* **2017**, *327*, 126–138. [CrossRef]

44. Deng, W.; Fergus, J.W. Effect of CMAS composition on hot corrosion behavior of gadolinium zirconate thermal barrier coating materials. *J. Electrochem. Soc.* **2017**, *164*, C526–C531. [CrossRef]

45. Song, W.; Lavallée, Y.; Hess, K.-U.; Kueppers, U.; Cimarelli, C.; Dingwell, D.B. Volcanic ash melting under conditions relevant to ash turbine interactions. *Nat. Commun.* **2016**, *7*, 10795. [CrossRef] [PubMed]

Article

Crack Healing in Mullite-Based EBC during Thermal Shock Cycle

Hyoung-IL Seo [1], Daejong Kim [2] and Kee Sung Lee [1],*

[1] School of Mechanical Engineering, Kookmin University, Seoul 02707, Korea; ssng3605@kookmin.ac.kr
[2] Advanced Materials Research Division, Korea Atomic Energy Research Institute, Daejeon 34057, Korea; dkim@kaeri.re.kr
* Correspondence: keeslee@kookmin.ac.kr

Received: 9 August 2019; Accepted: 10 September 2019; Published: 17 September 2019

Abstract: Crack healing phenomena were observed in mullite and mullite + Yb_2SiO_5 environmental barrier coating (EBC) materials during thermal shock cycles. Air plasma spray coating was used to deposit the EBC materials onto a Si bondcoat on a SiC_f/SiC composite substrate. This study reveals that unidirectional vertical cracks (mud cracks) formed after several thermal shock cycles; however, the cracks were stable for 5000 thermal shock cycles at a maximum temperature of 1350 °C. Moreover, the crack densities decreased with an increasing number of thermal shock cycles. After 3000 thermal shock cycles, cracks were healed via melting of a phase containing SiO_2 phase, which partially filled the gaps of the cracks and resulted in the precipitation of crystalline Al_2O_3 in the mullite. Post-indentation tests after thermal shock cycling indicated that the mullite-based EBC maintained its initial mechanical behavior compared to Y_2SiO_5. The indentation load–displacement tests revealed that, among the materials investigated in the present study, the mullite + Yb_2SiO_5 EBC demonstrated the best durability during repetitive thermal shocks.

Keywords: crack healing; environmental barrier coating; thermal shock; Hertzian indentations; mechanical behavior; CMC composite

1. Introduction

Gas turbines that operate at high temperatures are being developed for and installed in stationary and aviation applications. Higher operating temperatures can increase the energy efficiency and contribute to a cleaner environment. However, because materials that can withstand temperatures greater than 1000 °C are required, studies to develop a ceramic-based material beyond a single-crystalline superalloy are needed. In particular, all-ceramic materials are likely to be used for hot gas components such as combustor liners, blades, and nozzles, which require durable materials that can withstand high temperatures.

Silicon carbide exhibits both excellent heat resistance and high strength at high temperatures in addition to exceptional thermal shock resistance [1,2]. Silicon carbide also exhibits high hardness at room temperature and excellent wear resistance. The problem of brittleness can be overcome through the use of silicon carbide fibers (i.e., SiC-fiber-reinforced SiC (SiC_f/SiC) composites). Even if the silicon carbide matrix cracks, the fibers with relatively high mechanical strength support the mechanical load and suppress crack propagation, thereby preventing complete failure [3,4]. SiC_f/SiC composites are also expected to resist high-temperature creep and fatigue.

However, high-temperature oxidation and hot corrosion remain unsolved problems impeding the use of SiC_f/SiC composite materials in gas turbines. Because gas turbines operate in high-temperature environments for extended periods, the oxide film initially formed on their surface by the oxidation of silicon carbide is expected to function as a protective film that prevents further oxidation. However, in an environment where heating and cooling cycles are repeated, the entire system can be damaged because of inadequate protection by the oxide film. In particular, when exposed to a steam atmosphere at high temperature, SiC_f/SiC composite materials can undergo mass reduction due to high-temperature corrosion. This behavior is attributed to Si in a silicon carbide reacting with water vapor (H_2O) in the atmosphere to generate hydroxides such as $Si(OH)_x$ and $Si–O_x–H_y$. The hydroxide escapes into a gaseous state, possibly causing damage where the gas exits the matrix. Therefore, mass reduction due to high-temperature reactions with water vapor should be prevented to improve the operating lifetime of turbines [5,6].

To prevent the corrosion of Si-based ceramics at high temperatures, studies on the environmental barrier coatings (EBCs) have been extensively investigated [7–13]. The application of an oxide ceramic layer with a thickness of several tens to several hundreds of micrometers onto the silicon carbide surface can prevent high-temperature oxidation and corrosion by blocking the oxygen and steam penetration. Mullite and rare-earth silicates, which have excellent chemical resistance at high temperatures, are well known to be suitable materials for EBCs. Kang Lee first introduced a method to inhibit the reduction of mass of Si-based materials by applying an EBC [6]. This approach is feasible because the EBC exhibits sufficient thermal and chemical resistance and good mechanical properties. Specifically, the EBC is required to have a low thermal mismatch with the sublayer, high chemical resistance to steam, low oxygen permeability, good phase stability, and high impact and wear resistance. However, delamination of the coating layer occurs during multiple heating cycles of a gas turbine system, where heating and cooling are repeated. This delamination is due to differences in thermal expansion coefficients and elastic moduli between the coating and sublayer materials. As the heating and cooling are repeated, the expansion and contraction of the coating layer constrained by the sublayer material is restricted, inevitably resulting in the accumulation of stress in the coating layer. The literature reports numerous accounts of cracks forming in EBCs [14–16], where mud-cracks form in EBCs [17]; however, they do not grow to cause interface delamination because the cracks are unidirectional and vertical. Nonetheless, if these cracks can grow along the interface during continuous thermal cyclic loading, they may lead to interface delamination.

Crack healing of materials is an interesting subject. In particular, crack healing of brittle materials such as concrete or engineering ceramics is an interesting approach to prolonging their life [18–20]. In the case of concrete, crack healing occurs extrinsically. A crack healing agent in the form of filled capsules or particles is added to the concrete matrix, where the capsules or particles are ruptured upon crack penetration if cracking occurs. The healing agent is then released to fill and heal the crack. However, in the case of engineering ceramics, crack healing occurs intrinsically via a high-temperature reaction such as oxidation. For example, an additive material can react with atmospheric oxygen at a high temperature and fill the crack when it is exposed to high temperatures in air for a prolonged period. The initial strength has been reported to be recovered by crack healing [18,21,22].

In this study, mullite-based EBCs were applied by air plasma spray (APS) onto a Si bondcoat on a SiC_f/SiC composite. For comparison, a coating layer of Y_2SiO_5 was also prepared by APS. The prepared coating layer was subjected to repeated thermal shock cycles from 1350 °C to room temperature to investigate the effects of thermal shock cycles on crack density. The mechanical behavior of indentation load–displacement, as evaluated by Hertzian indentation, was subsequently analyzed to investigate the change in mechanical properties of the EBCs after the thermal shock cycles.

2. Experimental Procedures

2.1. Material Preparation

The starting commercial raw powders used in this study were Yb_2O_3 (3N, Kojundo Chemical Laboratory Co., Ltd., Saitama, Japan, submicron) or Y_2O_3 (5N, Hebei Pejin International Co., Ltd., Hebei, China, particle size of 3 μm) and SiO_2 (Kojundo Chemical Laboratory Co., Ltd., Saitama, Japan). The Yb_2SiO_5 or Y_2SiO_5 powder was mixed in molar composition of Yb_2O_3:SiO_2 = 1:1 or Y_2O_3:SiO_2 = 1:1, respectively, with zirconia balls in a ball mill. Each powder was dispersed in alcohol in the ball mill and mixed for 24 h. After mixing, they were dried in a fume hood for 24 h in air, for 4 h or more in a drying oven, and then sieved using a 60-μm sieve. The ball-milled powders were first heated at 600 °C for 1 h and then heated at 1400 °C for 20 h to synthesize Yb_2SiO_5 or Y_2SiO_5. The heat-treated powders were crushed and then sieved again.

The synthesized Yb_2SiO_5 powder was dispersed in alcohol with mullite powder (DURAMUL 325F, Washington Mills, NY, USA) and then mixed homogeneously in a ball mill with zirconia balls. The weight ratio between mullite and the synthesized Yb_2SiO_5 powder was 88:12. The mixed powders were subjected to the same conditions of drying and sieving previously described. The 100% mullite powder was also prepared from different mullite powders with an average diameter of 43 μm ($3Al_2O_3 \cdot SiO_2$, Yichang Kebo Refractories Co., Ltd., Yichang, China, KB-M-003).

The organic additives, which included a binder (PVA, Sigma-Aldrich, St. Louis, MO, USA) and a dispersant (San NOPCO, Tokyo, Japan, CERASRERSE 5468CF), were added to each synthesized or mixed powder with distilled water and zirconia balls in a ball mill for 24 h. At this time, 5 wt % of the binder and 4 wt % of the dispersant were added to the powder mixture.

The slurry was fed at a constant rate of 10 L/h and then sprayed as a soft agglomerated powder using the rotating atomizer of a spray dryer (Sewon Hardfacing Co., Mokpo, Korea). At this time, the inlet temperature was controlled at 170–210 °C, the outlet temperature at 70–110 °C, and the rotation speed of the atomizer at 6000 to 10,000 rpm.

Heat treatment was carried out at 600–800 °C in air to remove the organic additives in the spherical agglomerates. The agglomerates were continuously heat treated at 1250 °C for 2 h to impart strength to the powders in air. The heat-treated powders were sieved to various sizes and then granules within the range of 10 to 63 μm were classified and used as a powder for coating.

Liquid silicon infiltration (LSI) SiC-fiber-reinforced SiC composite material was prepared at the Korea Institute of Energy Research by reactive sintering after Si liquid infiltration with a diameter of 1 inch (25.4 mm) and a thickness of approximately 3 mm as a substrate for the EBC. The preforms were prepared by laminating silicon carbide fibers (Ube Industry, SA grade, Japan) with [0°, 90°] stacking and then filling them with phenolic resin to prepare fiber-reinforced plastics. The fiber fraction was maintained at 40 vol %. The prepared preform was cured at a constant temperature of 180 °C while being compressed through vacuum bagging. Afterwards, the cured product was subjected to carbonization of the phenolic resin at 1000 °C under a nitrogen gas atmosphere. The SiC-fiber-reinforced composite material was then prepared by an LSI process in which elemental silicon was heated to 1450 °C in a vacuum atmosphere to melt and infiltrate.

Silicon bondcoat having a thermal expansion coefficient intermediate between the coating material and the substrate material was prepared from Si powder (Si, particle size 40 μm, Saint Gobain, Anderlecht, Belgium).

Sand blasting was performed with alumina sand to improve bonding strength between the substrate and the bondcoat and also between the bondcoat and the topcoat. The bondcoat material was coated with a high-speed flame coating method (HVOF) to a thickness of approximately 400 μm, and a topcoat was applied by APS (9 MB, Sulzer Metco Holding AG, Winterthur, Switzerland) to a thickness of approximately 400 μm.

Mullite, mullite + Yb_2SiO_5, and Y_2SiO_5, prepared as previously described, were coated on the bondcoat. The coating powders were supplied to the plasma stream at a constant feed rate of 10 g/min with a mixed gas (mixing ratio of Ar/He = 55/5), and the current and voltage were kept constant at 500–530 A and 99 V, respectively. The distance between the spray gun and the substrate to be coated was controlled to 100–120 mm, and the sprayed velocity was controlled at a rate of 1000 mm/s to obtain a coating layer of constant thickness.

2.2. Thermal/Mechanical Characterization

The surface of the coating layer was ground and then polished with 15, 6, 3 and 1 µm diamond suspensions to observe the cracks in the EBCs. The thickness was measured at each polishing step so that the degree of polishing was controlled to be the same for all samples.

Thermal shock tests were performed to evaluate the durability of the EBC against repetitive thermal fatigue. The thermal shock was induced using a laboratory-made horizontal thermal shock machine in which the chamber could be moved horizontally by a motor. The schematic diagram of the thermal shock test is shown in Figure 1a. The test samples were placed on a fixed hotplate such that the substrate was positioned to receive heat from the lower part of the hot plate, and the coating layer was exposed to the top to receive heat from the chamber. First, the test samples placed on the hotplate were maintained at a temperature of 1100 and 1350 °C in the upper coating layer after the heating chamber was closed; the heating rate was 5 °C/min. After the target temperature had been maintained for 10 min, the upper chamber was moved to expose the sample to room temperature for 4 min as shown in Figure 1b, and then the chamber was moved back to expose the sample at 1350 °C again. This cycle was defined as one cycle, and repetitive thermal shocks were applied for a maximum of 5000 cycles.

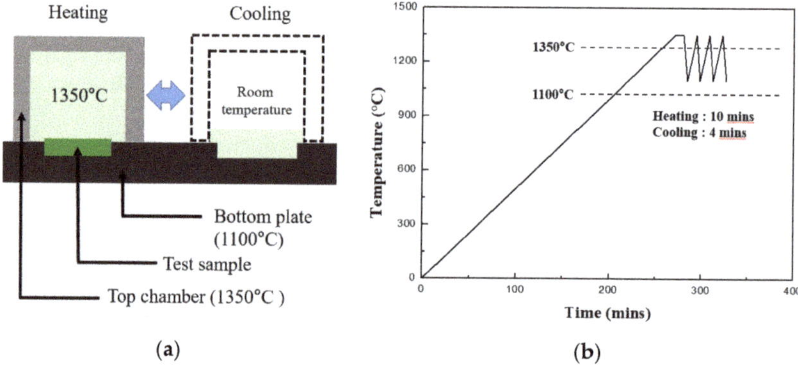

Figure 1. (a) Schematic of the thermal shock test from 1350 °C to room temperature; (b) temperature profiles during the thermal shock tests in this study.

We observed the surface using an optical microscope (Olympus, GX51, Tokyo, Japan) after thermal shock tests of 1000, 2000, 3000, 4000, and 5000 cycles. Cracks and damages were observed with an optical microscope, and the density of cracks was measured with an image analyzer. For this purpose, the surface was observed at 200× magnification using an optical microscope. The crack density was defined by measuring the number of cracks intercepting the baseline after the baseline had been marked at positions of 1/4, 1/2, and 3/4 of each line of the horizontal, vertical, and diagonal length, respectively. We defined the crack density as the number of cracks divided by the length of the baseline and expressed it in terms of the number per inch. The numbers of cracks were determined from 10 pictures, and the mean values were plotted. The crack morphology of the surface was also observed with a high-resolution optical microscope (VHX-5000, Keyence Co., Itasca, IL, USA) after thermal shock tests of 100 and 5000 cycles.

Scanning electron microscopy (SEM) was used to observe the cracks in detail after crack healing. The electron microscope (JSM-6010F, JEOL Co., Tokyo, Japan) was operated at an acceleration voltage of 10 kV. Energy-dispersive X-ray spectroscopy (EDS) was used to determine the element species with an acceleration voltage of 10 kV and a spot size of 3.0.

The phase of the coating layer was analyzed by X-ray diffraction (XRD, Rigaku Co., Ltd., Tokyo, Japan) to analyze the change in the phase before and after the crack healing. The XRD analyses were conducted with Cu Kα radiation generated at 45 kV and 200 mA with a scan rate of 2°/min over the 2θ range from 20° to 80°.

Spherical indentations were induced to compare and evaluate the mechanical behavior before and after the thermal shock test. The schematic diagram of spherical indentation is shown in Figure 2. Indentation load–displacement curves were obtained during indentation tests. A tungsten carbide ball with a radius of r = 3.18 mm was attached to the jig of a universal testing system (Instron 5567, Boston, MA, USA) and the pre-load was applied to the coating layer to P = 10 N at a rate of 3 N/min. The load was then applied to P = 500 N at a rate of 0.1 mm/min. The load was removed at the same rate as soon as the load had reached P = 500 N. The displacement was measured with an extensometer during load changes. Tests were conducted on more than five areas of the coating layer of the test specimens, and the indentation load and displacement results were averaged and plotted. The relative hardness was calculated from relative displacement after unloading in each indentation load–displacement and compared [23–26].

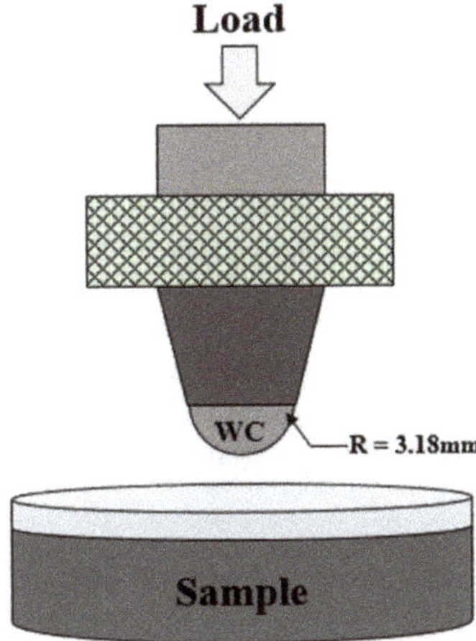

Figure 2. Schematic of a spherical indentation using a WC ball with a radius of 3.18 mm in this study.

3. Results and Discussion

The cracks formed on the surface and cross section of the EBC coating layer by thermal shock were observed by optical microscopy, as shown in Figure 3. Thermal shock tests were carried out at 1350 °C on the coating surface of the sample and at 1100 °C on the bottom of the sample because a gas turbine operating at 1600 °C is cooled simultaneously by the cooling gas through the flow path. Mud-type cracks formed on the surface. After 100 thermal shock cycles, cracks developed in the EBC layer fabricated by APS, as shown in Figure 3a. We considered that these cracks developed because of the difference in thermal expansion coefficients between the EBC coating and the metal sublayer. Several researchers have observed and reported mud cracks in EBCs [8]. However, cross-sectional observations of these cracks reveal that they formed in a direction perpendicular to the interface, not parallel to the interface. The vertical cracks shown in Figure 3b propagated through 400 μm of thickness; their propagation stopped at the top of the bondcoat layer. Such cracks can be stable if they do not induce horizontal cracking along the interface between the EBC and the sublayer, but rather tolerate strain caused by stress, thereby contributing to preventing delamination of the coating layer [27,28].

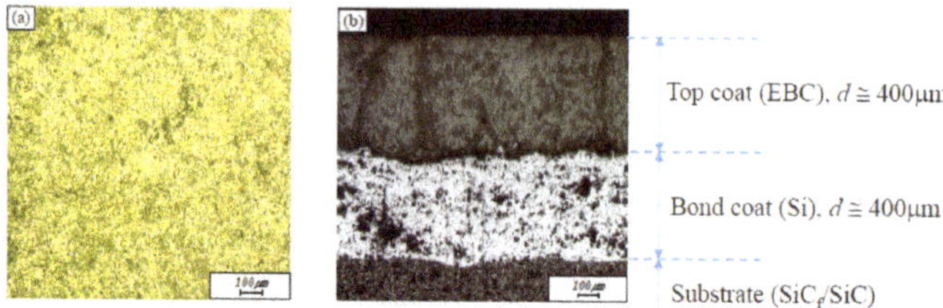

Figure 3. Optical micrographs of the (**a**) surface and (**b**) cross section of EBCs after 100 thermal shock cycles at 1350 °C.

The results of the measurements of the crack density developed on the surface of the EBC by thermal shock are shown in Figure 4. The periodic repetition of heating and cooling in turbine systems, and especially an abrupt stop of the system, will exacerbate crack formation. Therefore, the stability of cracks was evaluated by thermal shock tests. The mullite-based EBCs (i.e., the mullite and mullite + Yb$_2$SiO$_5$ EBCs) were not delaminated when the thermal shock cycle of exposure to room temperature was repeated for 5000 cycles while the coating material was maintained at 1350 °C and the SiC$_f$/SiC substrate material was maintained at 1100 °C. Even though a crack is generated initially, interface delamination does not occur because vertical cracks are formed. That is, an initially developed crack in the EBC layer is very stable. Although the Y$_2$SiO$_5$ EBC was stable for 4000 cycles, interface delamination occurred after 5000 thermal shock cycles. The crack density on the surface of a sample after each 1000 cycles in the thermal shock test was measured and plotted in Figure 4. As shown in the graph, the changes in crack density depend on the EBC material. The crack density did not substantially change compared with the initial crack density in the case of Y$_2$SiO$_5$, whereas the crack density of the mullite-based EBC increased slightly and then drastically decreased after 3000 cycles. Furthermore, it exhibited a substantial decrease in crack density after 5000 cycles.

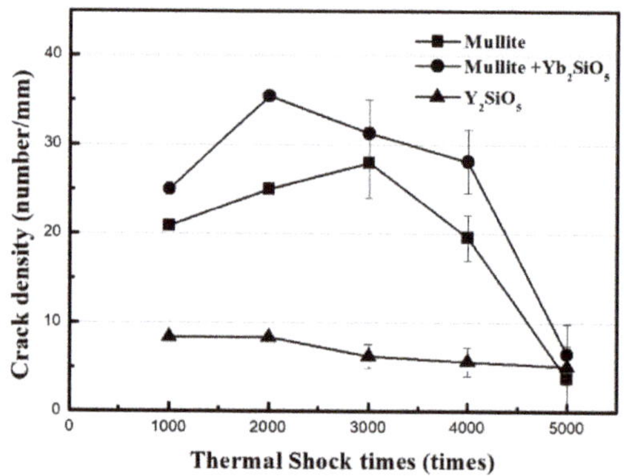

Figure 4. Plot of crack density of EBCs after each 1000 cycles of the thermal shock test at 1350 °C.

The optical microscope images in Figure 5 show the surfaces of the mullite, mullite + Yb$_2$SiO$_5$, and Y$_2$SiO$_5$ after 2000 and 4000 cycles of thermal shock, indicating crack healing. Although obvious cracks are observed after 2000 cycles of thermal shock irrespective of the EBC material (Figure 5a–c), surprisingly, the surface cracks were not observed after 4000 cycles of thermal shock (Figure 5d–f). In the case of Y$_2$SiO$_5$, relatively large macrocracks developed and were maintained without the formation of microcracks; thus, the crack density did not substantially change. By contrast, the population of microscale cracks increased at 2000 cycles and new phases covered the surface completely at 4000 cycles in the mullite-based EBC.

Figure 5. Optical micrographs of the surface of the EBCs after the thermal shock test at 1350 °C:
(**a**) mullite, 2000 cycles; (**b**) mullite + Yb_2SiO_5, 2000 cycles; (**c**) Y_2SiO_5, 2000 cycles; (**d**) mullite, 4000 cycles;
(**e**) mullite + Yb_2SiO_5, 4000 cycles; and (**f**) Y_2SiO_5, 4000 cycles.

The cracks on the surface of the mullite-based EBC after 5000 cycles of thermal shock were observed with a high-resolution optical microscope, as shown in Figure 6. The optical microscope images of the surfaces of the mullite-based EBC after 100 cycles are also presented in the figure. Although mud cracks were noticeable after 100 cycles of thermal shock, the crack gaps were clearly filled after 5000 cycles of thermal shock.

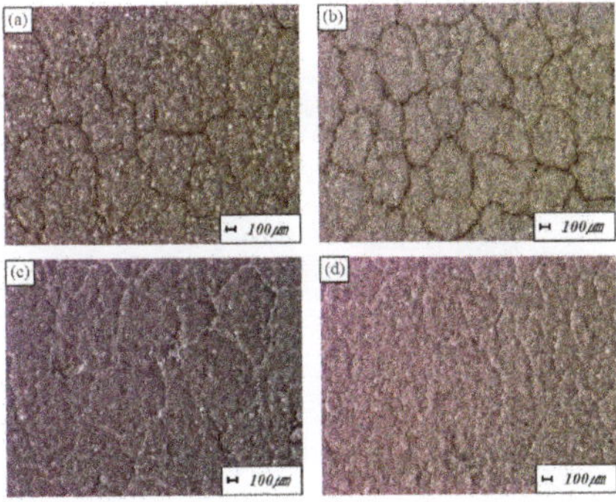

Figure 6. High-resolution optical micrographs of the surface of the EBCs after the thermal shock test at 1350 °C: (**a**) mullite, 100 cycles; (**b**) mullite + Yb_2SiO_5, 100 cycles; (**c**) mullite, 5000 cycles; and (**d**) mullite + Yb_2SiO_5, 5000 cycles.

Figure 7 shows the cracks of the mullite-based EBC after 100 and 5000 thermal shock cycles, as observed at high magnification by SEM. Only cracks and grains are observed after 100 cycles of thermal shock; cracks are observed around the grain. By contrast, after 5000 thermal shock cycles, new phases are formed to fill the gaps and cover the cracks. The results of the EDS composition analysis of these phases indicate that they contain only Si and Al, which are the constituents of mullite. We therefore speculate that the mullite itself underwent a change when repeatedly exposed to high temperatures.

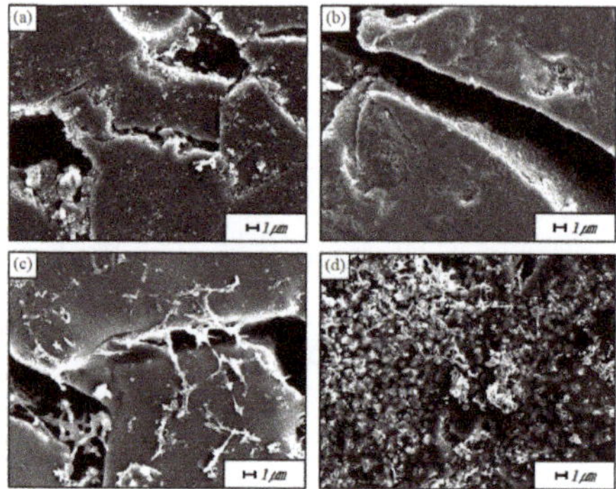

Figure 7. SEM micrographs showing the microstructure on the surface of the EBC after thermal shock test at 1350 °C: (**a**) mullite, 100 cycles; (**b**) mullite + Yb$_2$SiO$_5$, 100 cycles; (**c**) mullite, 5000 cycles; (**d**) mullite + Yb$_2$SiO$_5$, 5000 cycles.

Figure 8 shows the cracks on the cross-section of the mullite, mullite + Yb$_2$SiO$_5$, and Y$_2$SiO$_5$ after 5000 thermal shock cycles, as observed by SEM. Each arrow in the mullite, mullite + Yb$_2$SiO$_5$ indicate crack healed region. On the other hand, each arrow in the Y$_2$SiO$_5$ indicate cracked region. Only cracks are observed after 5000 cycles of thermal shock in the case of Y$_2$SiO$_5$ (Figure 8c). By contrast, after 5000 thermal shock cycles, new phases are formed to fill the gaps and cover the cracks, and this result is more pronounced in the mullite + Yb$_2$SiO$_5$ (Figure 8b).

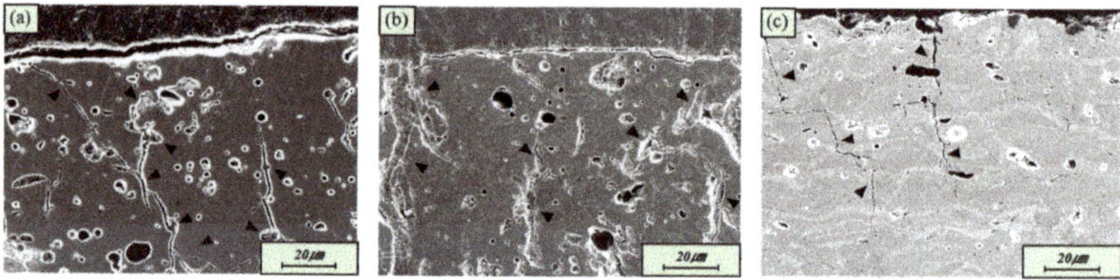

Figure 8. SEM micrographs showing the microstructure on the cross-section of the EB after thermal shock test at 1350 °C: (**a**) mullite, 5000 cycles; (**b**) mullite + Yb$_2$SiO$_5$, 5000 cycles; and (**c**) Y$_2$SiO$_5$, 5000 cycles.

XRD analysis was performed on the EBC materials after 5000 thermal shock cycles for the analysis of the phases formed around the cracks, as shown in Figure 9. In the case of the mullite material, an Al$_2$O$_3$ phase was detected in addition to the mullite phase. Therefore, the newly formed phases in Figures 7 and 8 are attributed to mullite that melted under repeated thermal shock cycles at high temperature. We speculate that the low-melting-point phases containing SiO$_2$ migrated into the cracks, filling the gaps of the cracks, and that the Al$_2$O$_3$ phases precipitated as crystalline phase during cooling. In particular, Al$_2$O$_3$ phases are considered desirable because they exhibit excellent oxidation and corrosion resistance at high temperatures and can therefore resist oxidation and corrosion of EBC materials after crack healing. Similarly, an Al$_2$O$_3$ phase was detected along with the mullite, Yb$_2$SiO$_5$, and Yb$_2$Si$_2$O$_7$ phases in the case of the mullite + Yb$_2$SiO$_5$ material, as shown in Figure 9b. However, in the case of Y$_2$SiO$_5$, Y$_2$O$_3$, and SiO$_2$ phases were detected along with Y$_2$SiO$_5$ and Y$_2$Si$_2$O$_7$ phases, as shown in Figure 9c. The Y$_2$SiO$_5$ phase melts at high temperatures and is decomposed into crystalline Y$_2$O$_3$ and SiO$_2$, similar to the process that occurs in the mullite-based EBC.

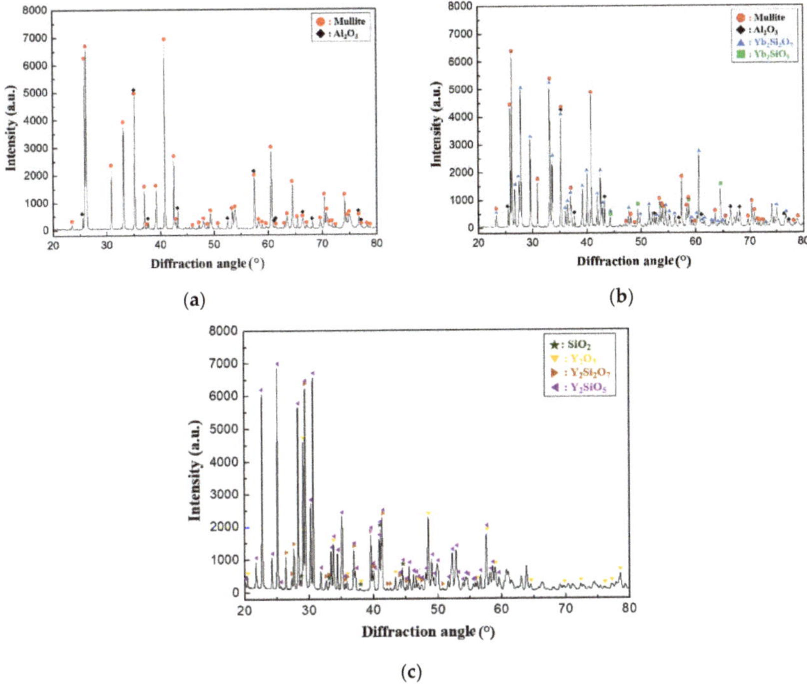

Figure 9. XRD patterns of (**a**) mullite, (**b**) mullite + Yb$_2$SiO$_5$, and (**c**) Y$_2$SiO$_5$ after 5000 thermal shock cycles at 1350 °C.

Rare-earth silicate materials such as Y$_2$SiO$_5$ and Yb$_2$SiO$_5$ have been reported to exhibit less mass loss than mullite in a steam atmosphere at relatively higher temperatures [11,17,29,30]. However, if cracks occur in the coating layer because of thermal shock, the crack healing ability of Y$_2$SiO$_5$ and Yb$_2$SiO$_5$ is considered to be relatively poor. Therefore, studies are needed to develop a eutectic phase or new crack healing agent. The mullite + Yb$_2$SiO$_5$ material suggested in this study exhibits high water vapor resistance because of the Yb$_2$SiO$_5$ phase and exhibits crack healing behavior because of the mullite phase. Therefore, the mullite + Yb$_2$SiO$_5$ can be used as a high-temperature EBC material with crack healing ability.

The surfaces of the EBC materials were contacted with a WC ball with a radius of 3.18 mm to a load P of 500 N and then unloaded to determine the mechanical behavior after crack healing, as shown in Figure 10. The graphs show the test results for the sample before the thermal shock and the sample after 5000 thermal shock cycles. The Y$_2$SiO$_5$ EBC shows a large change from elastic to plastic after 5000 thermal shock cycles, as shown in Figure 10c because the test was carried out on the delaminated coating samples. By contrast, the load–displacement behavior of the mullite-based EBCs, which did not exhibit interfacial delamination even after 5000 thermal shock cycles, did not substantially differ from their initial indentation load–displacement behavior, as shown in Figure 10a,b. In the case of mullite, the residual displacement related to the hardness of the material was slightly increased after thermal shock, and the tangential slope of the stress–strain at unloading, which is related to the modulus of elasticity, decreased after thermal shock [10,25,26,31]. Thus, the hardness and elastic modulus of the sample decreased by 5000 thermal shock cycles. The decrease in the mechanical properties is attributed to cracks even though the sintering of the material during repetitive exposure to high temperatures occurs [32,33]. However, the hardness and elastic modulus of the mullite + Yb$_2$SiO$_5$ did not substantially change even after 5000 thermal shock cycles. Therefore, the initial mechanical properties of the mullite + Yb$_2$SiO$_5$ were recovered by crack healing.

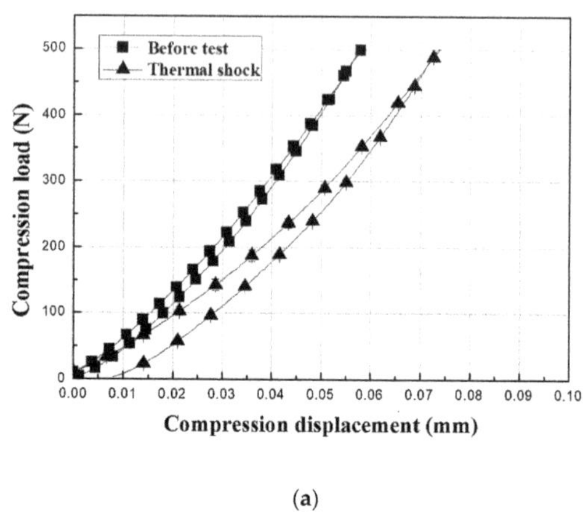

(**a**)

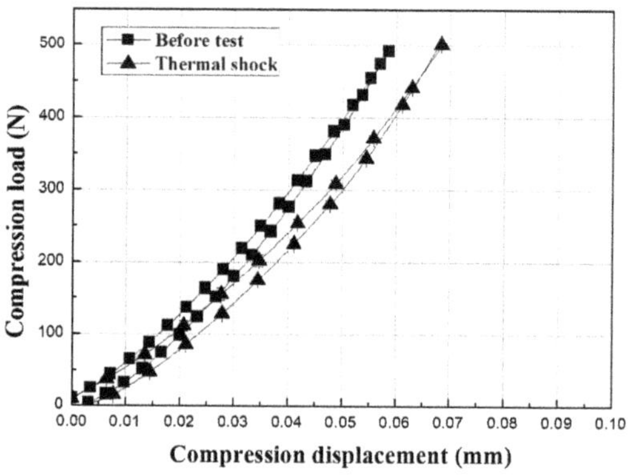

(**b**)

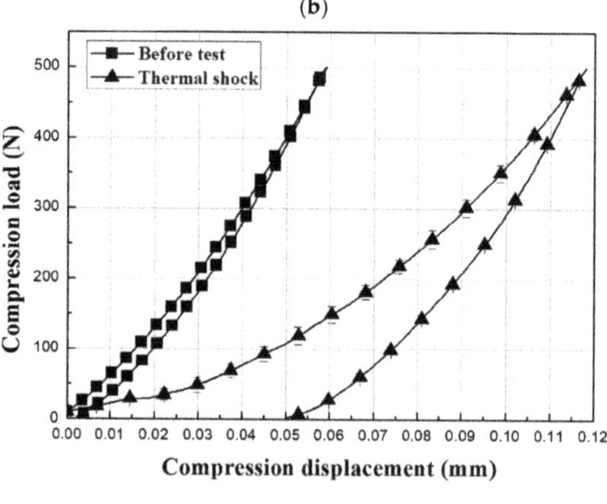

(**c**)

Figure 10. Plot of indentation load–displacement curves of (**a**) mullite, (**b**) mullite + Yb_2SiO_5 and (**c**) Y_2SiO_5 using a WC ball with a radius of 3.18 mm at load P = 500 N.

Figure 11 shows the change in the relative hardness of the material, as revealed by analysis of the residual displacement of the indentation load–displacement curves in Figure 10. The hardness value before the thermal shock test is assumed to be 100%, and the hardness value after 5000 cycles of thermal shock is plotted in percent. As shown in the graph, the mullite + Yb_2SiO_5 maintains its initial hardness value. These results indicate that the mullite + Yb_2SiO_5 is a potential EBC material because gas turbine coating materials exposed to repeated thermal cycles must not only not exhibit interfacial delamination due to the thermal shock but also resist impact and wear due to external FOD.

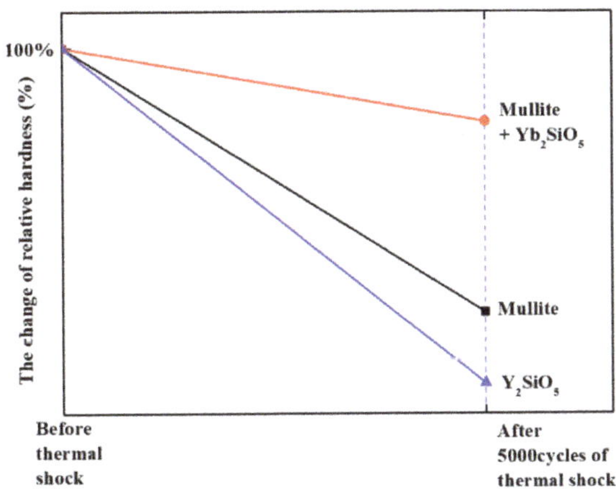

Figure 11. Hardness changes of EBC materials during thermal shock tests.

4. Conclusions

The purpose of this study was to introduce an EBC to protect SiC_f/SiC composites from high-temperature oxidation and corrosion and to investigate the durability of the coating when exposed to high temperatures and repeated thermal shocks. Crack formation, crack stability, and crack healing were investigated after specimens were subjected to repeated heating to 1350 °C followed by cooling. In addition, spherical post-indentation tests were performed to determine the change in mechanical behavior after thermal shock. As a result, the following conclusions were obtained:

(1)   EBC layers composed of mullite, mullite + Yb_2SiO_5, and Y_2SiO_5 were coated via APS onto a Si bondcoat. The bondcoat was previously coated onto a SiC_f/SiC composite. All of the coating process conditions were controlled to achieve a uniform thickness of 400 μm. In addition, the prepared EBC coating layers were expected to exhibit less than 10% porosity to prevent water vapor penetration at high temperatures.

(2)   Mud cracks developed on the surface and unidirectional vertical cracks formed on the cross section after several thermal shock cycles in the EBC layer fabricated by APS. The thermal shock cycles involved repeated exposure to room temperature after the coating material was maintained at 1350 °C and the SiC_f/SiC substrate material was maintained at 1100 °C. However, the cracks in the mullite-based EBC were very stable and did not lead to interfacial delamination until 5000 thermal shock cycles were completed. This behavior was attributed to the advantage of strain tolerance due to vertical cracks. The Y_2SiO_5 coating material also exhibited good stability: interface delamination did not occur until 4000 thermal shock cycles were completed.

(3)   Crack healing phenomena were observed in the present study when the thermal shock cycles were repeated for 4000–5000 cycles for the mullite-based EBCs, where the temperature of the EBC coating layer was maintained at 1350 °C and the SiC_f/SiC composite material was maintained at 1100 °C. The cracks on the cross-section as well as surface were covered with a new phase in the case of the mullite-based EBC. Microscopic observations revealed that a new phase was formed

around the gap of the cracks. A crystalline phase of Al_2O_3 was detected by XRD analysis. We considered that the mullite phase melted at high temperatures and filled the gaps of the cracks and that a crystalline phase of Al_2O_3 was precipitated during cooling, resulting in crack healing. Crystalline SiO_2 and Y_2O_3 phases were observed in the case of Y_2SiO_5.

(4) The mechanical indentation load–displacement behaviors of the EBC samples were characterized using a spherical indentation method after the specimens were subjected to 5000 thermal shock cycles. The mullite-based EBC showed no substantial change in mechanical behavior even after 5000 thermal shock cycles, whereas the Y_2SiO_5 exhibited a change from elastic to plastic behavior because of interfacial delamination. In particular, the mullite + Yb_2SiO_5 EBC maintained its initial mechanical behavior even after 5000 thermal shock cycles because of its crack healing behavior.

Author Contributions: Conceptualization, K.S.L., H.S.; Methodology, H.S., D.K.; Software, D.K., H.S.; Validation, D.K.; Formal Analysis, H.S.; Investigation, H.S.; Resources, K.S.L.; Data Curation H.S., D.K.; Writing—Original Draft Preparation, K.S.L., H.S.; Writing—Review and Editing, K.S.L., H.S. and D.K.; Visualization, K.S.L., D.K.; Supervision, K.S.L.; Project Administration, K.S.L.; Funding Acquisition, K.S.L., D.K.

Funding: This research was funded by the IT R&D program of MOTIE/KEIT [NO. 20000192, Development on the crack self-healing technology of CMC composite and coating material for gas turbine hot gas component].

Acknowledgments: We wish to thanks to Sewon Hardfacing Company and KIER for coating and the substrate preparation.

Conflicts of Interest: The authors declare no conflict of interest.

References

1. Lawn, B.R.; Padture, N.P.; Cai, H.; Guiberteau, F. Making ceramics "ductile". *Science* **1994**, *263*, 1114–1116. [CrossRef]

2. Kim, Y.W.; Jang, S.H.; Nishimura, T.; Choi, S.Y.; Kim, S.D. Microstructure and high temperature strength of silicon carbide with 2000ppm yttria. *J. Eur. Ceram. Soc.* **2017**, *37*, 4449–4455. [CrossRef]

3. Yoshida, K.; Budiyanto; Imai, M.; Yano, T. Processing and microstructure of silicon carbide fiber-reinforced silicon carbide composite by hot-pressing. *J. Nucl. Mater.* **1998**, *258–263*, 1960–1965. [CrossRef]

4. Fantozzi, G.; Reynaud, P. Mechanical behaviour of SiC fiber-reinforced ceramic matrix composites. *Compr. Hard Mater.* **2014**, *2*, 345–366.

5. Eaton, H.E.; Linsey, G.D.; More, K.L.; Kimmel, J.B.; Price, J.R.; Miriyala, N. EBC protection of SiC/SiC composites in the gas turbine combustion environment. *ASME* **2000**, *4*, V004T02A010.

6. Lee, K.N. Current status of environmental barrier coatings for Si-based ceramics. *Surf. Coat. Technol.* **2000**, *133–134*, 1–7. [CrossRef]

7. Harder, B.J.; Faber, K.T. Transformation kinetics in plasma-sprayed barium- and strontium-doped aluminosilicate (BSAS). *Scr. Mater.* **2010**, *62*, 282–285. [CrossRef]

8. Girolamo, G.D.; Blasi, C.; Pilloni, L.; Schioppa, M. Microstructure and thermal properties of plasma sprayed mullite coatings. *Ceram. Int.* **2010**, *36*, 1389–1395. [CrossRef]

9. Jang, B.-K.; Feng, F.J.; Lee, K.S.; Garcia, E.; Nistal, A.; Nagashima, N.; Kim, S.; Oh, Y.-S.; Kim, H.-T. Thermal behavior and mechanical properties of Y_2SiO_5 environmental barrier coatings after isothermal heat treatment. *Surf. Coat. Technol.* **2016**, *308*, 24–30. [CrossRef]

10. Feng, F.J.; Jang, B.-K.; Park, J.Y.; Lee, K.S. Effect of Yb_2SiO_5 addition on the physical and mechanical properties of sintered Mullite ceramic as an environmental barrier coating material. *Ceram. Int.* **2016**, *42*, 15203–15208. [CrossRef]

11. Tian, Z.; Zheng, L.; Li, Z.; Li, J.; Wang, J. Exploration of the low thermal conductivities of γ-$Y_2Si_2O_7$, β-$Y_2Si_2O_7$, β-$Yb_2Si_2O_7$, and β-$Lu_2Si_2O_7$ as novel environmental barrier coating candidates. *J. Eur. Ceram. Soc.* **2016**, *36*, 2813–2823. [CrossRef]

12. Ueno, S.; Ohji, T.; Lin, H.-T. Recession behavior of Lu_2SiO_5 under a high speed steam jet at high temperatures. *Ceram. Int.* **2011**, *37*, 1185–1189. [CrossRef]

13. Liu, J.; Zhang, L.; Yang, J.; Cheng, L.; Wang, Y. Fabrication of SiCN-$Sc_2Si_2O_7$ coatings on C/SiC composites at low temperatures. *J. Eur. Ceram. Soc.* **2012**, *32*, 705–710. [CrossRef]

14. Liu, J.; Zhang, L.; Liu, Q.; Cheng, L.; Wang, Y. Structure design and fabrication of environmental barrier coatings for crack resistance. *J. Eur. Ceram. Soc.* **2014**, *34*, 2005–2012. [CrossRef]

15. Mesquita-Guimaraes, J.; Garcia, E.; Osendi, M.I.; Sevecek, O.; Bermejo, R. Effect of againg on the onset of cracks due to redistribution of residual stresses in functionally graded environmental barrier coatings. *Compos. Part B* **2014**, *61*, 199–205. [CrossRef]

16. Chen, Y.; Lu, Y.; Ye, Q.; Wang, Y. A Self-healing environmental barrier coating: $TiSi_2$-doped $Y_2Si_2O_7$/barium strontium aluminosilicate coating. *Surf. Coat. Technol.* **2016**, *307*, 436–440. [CrossRef]

17. Richards, B.T.; Sehr, S.; Franqueville, F.; Begley, M.R.; Wadley, H.N.G. Fracture mechanisms of ytterbium monosilicate environmental barrier coatings during cyclic thermal exposure. *Acta Mater.* **2016**, *103*, 448–460. [CrossRef]

18. Ando, K.; Furusawa, K.; Takahashi, K.; Sato, S. Crack-healing ability of structural ceramics and a new methodology to guarantee the structural integrity using the ability and proof-test. *J. Eur. Ceram. Soc.* **2005**, *25*, 549–558. [CrossRef]

19. Tittelboom, K.V.; Belie, N.D. Self-healing in cementitious materials—A review. *Materials* **2013**, *6*, 2182–2217. [CrossRef]

20. Nozahi, F.; Estournes, C.; Carabat, A.L.; Sloof, W.G.; Zwaag, S.; Monceau, D. Self-healing thermal barrier coating systems fabricated by spark plasma sintering. *Mater. Des.* **2018**, *143*, 204–213. [CrossRef]

21. Ando, K.; Chu, M.C.; Tsuji, K.; Hirasawa, T.; Kobayashi, Y.; Sato, S. Crack healing behaviour an d high-temperature strength of mullite/SiC composite ceramic. *J. Eur. Ceram. Soc.* **2002**, *22*, 1313–1319. [CrossRef]

22. Lee, S.K.; Ishida, W.; Lee, S.Y.; Nam, K.W.; Ando, K. Crack-healing behavior and resultant strength properties of silicon carbide ceramics. *J. Eur. Ceram. Soc.* **2005**, *25*, 569–576. [CrossRef]

23. Lee, D.H.; Kang, N.K.; Lee, K.S.; Moon, H.S.; Kim, H.T.; Kim, C. Evaluation of thermal durability of thermal barrier coating and change in mechanical behavior. *J. Korean Ceram. Soc.* **2017**, *54*, 314–322. [CrossRef]

24. Hutchings, I.M. The contributions of David Tabor to the science of indentation hardness. *J. Mater. Res.* **2009**, *24*, 581–589. [CrossRef]

25. Fischer-Cripps, A.C. Critical review of analysis and interpretation of nanoindentation test data. *Surf. Coat. Technol.* **2006**, *200*, 4153–4165. [CrossRef]

26. Fischer-Cripps, A.C. A review of analysis methods for sub-micron indentation testing. *Vacuum* **2000**, *58*, 569–585. [CrossRef]

27. Jadhav, A.; Padture, N.P.; Wu, F.; Jordan, E.H.; Gell, M. Thick ceramic thermal barrier coatings with high durability deposited using solution-precursor plasma spray. *Mater. Sci. Eng. A* **2005**, *405*, 313–320. [CrossRef]

28. Guo, H.B.; Kuroda, S.; Murakami, H. Segmented thermal barrier coatings produced by atmospheric plasma spraying hollow powders. *Thin Solid Film.* **2006**, *506*, 136–139. [CrossRef]

29. Lee, K.N.; Fox, D.S.; Bansal, N.P. Rare earth silicate environmental barrier coatings for SiC/SiC composites and Si_3N_4 ceramics. *J. Eur. Ceram. Soc.* **2005**, *25*, 1705–1715. [CrossRef]

30. Fritsch, M.; Klemm, H. The water vapor hot gas corrosion behaviour of Al_2O_3-Y_2O_3 materials, Y_2SiO_5 and Y_3AlO_{12}-coated alumina in combustion environment. In Proceedings of the 30th International Conference and Exposition on Advanced Ceramics and Composites, Cocoa Beach, FL, USA, 25–30 January 2006.

31. Oliver, W.C.; Pharr, G.M. Measurement of hardness and elastic modulus by instrumented indentation: Advances in understanding and refinements to methodology. *J. Mater. Res.* **2004**, *19*, 3–20. [CrossRef]

32. Chen, C.; Guo, H.; Gong, S.; Zhao, X.; Xiao, P. Sintering of electron beam physical vapor deposited thermal barrier coatings under flame shock. *Ceram. Int.* **2013**, *39*, 5093–5102. [CrossRef]

33. Fleck, N.A.; Cocks, A.C.F.; Lampenscherf, S. Thermal shock resistance of air plasma sprayed thermal barrier coatings. *J. Eur. Ceram. Soc.* **2014**, *34*, 2687–2694. [CrossRef]

Article

Crack Initiation Criteria in EBC under Thermal Stress

Emi Kawai [1], Hideki Kakisawa [2], Atsushi Kubo [3], Norio Yamaguchi [1], Taishi Yokoi [4], Takashi Akatsu [5], Satoshi Kitaoka [1] and Yoshitaka Umeno [3,*]

[1] Japan Fine Ceramics Center, Nagoya 456-8587, Japan; emi_kawai@jfcc.or.jp (E.K.); yamaguchi@jfcc.or.jp (N.Y.); kitaoka@jfcc.or.jp (S.K.)
[2] National Institute for Materials Science, Ibaraki 305-0047, Japan; KAKISAWA.Hideki@nims.go.jp
[3] Institute of Industrial Science, The University of Tokyo, Tokyo 153-8505, Japan; kubo@ulab.iis.u-tokyo.ac.jp
[4] Institute of Biomaterials and Bioengineering, Tokyo Medical and Dental University, Tokyo 101-0062, Japan; yokoi.taishi.bcr@tmd.ac.jp
[5] Faculty of Art and Regional Design, Saga University, Saga 840-8502, Japan; akatsu@cc.saga-u.ac.jp
* Correspondence: umeno@iis.u-tokyo.ac.jp; Tel.: +81-3-5452-6902

Received: 12 September 2019; Accepted: 21 October 2019; Published: 24 October 2019

Abstract: For design of multi-layered environmental barrier coatings (EBCs), it is essential to assure mechanical reliability against interface crack initiation and propagation induced by thermal stress owing to a misfit of the coefficients of thermal expansion between the coating layers and SiC/SiC substrate. We conducted finite element method (FEM) analyses to evaluate energy release rate (ERR) for interface cracks and performed experiment to obtain interface fracture toughness to assess mechanical reliability of an EBC with a function of thermal barrier (T/EBC; SiC/SiAlON/mullite/Yb-silicate gradient composition layer/Yb$_2$SiO$_5$ with porous segment structure) on an SiC/SiC substrate under thermal stress due to cooling in fabrication process. Our FEM analysis revealed that a thinner SiAlON layer and a thicker mullite layer are most suitable to reduce ERRs for crack initiation at the SiC/SiAlON, SiAlON/mullite and mullite/Yb$_2$Si$_2$O$_7$ interfaces. Interface fracture tests of the T/EBC with layer thicknesses within the proposed range exhibited fracture at the SiC/SiAlON and SiAlON/mullite interfaces. We also estimated the approximate fracture toughness for the SiC/SiAlON and SiAlON/mullite interfaces and lower limit of fracture toughness for the mullite/Yb$_2$Si$_2$O$_7$ interface. Comparison between ERR and fracture toughness indicates that the fabricated T/EBC possesses sufficient mechanical reliability against interface crack initiation and propagation.

Keywords: fabrication process; layer thickness design; interface crack initiation/propagation; fracture toughness; energy release rate; finite element method analysis

---

1. Introduction

Silicon carbide (SiC) fiber reinforced SiC matrix (SiC/SiC) composites are one-third lighter and have approximate 100–200 K higher heat resistance than current heat-resistant super alloys (Nickel-based super alloys) [1–3]. Application of SiC/SiC to advanced hot-section components in airplane engines is expected for improving fuel consumption and curbing emission of carbon dioxide [2,3]. SiC/SiC composites react with oxygen to form silica, which hinders degradation of the composites at high temperature over 1373–1473 K [4,5]. In the condition of high temperature and humidity (i.e., combustion gas environment), however, the silica reacts with water vapor to form silicic acid (Si(OH)$_4$)) gas [4,5], leading to wall thinning of the composites. Thus, application of the SiC/SiC composites to the hot-section components inevitably requires environmental barrier coatings (EBCs).

To achieve superior environmental shielding performance and thermomechanical durability, EBCs consist of several layers with different shielding functions [6,7]. For using EBCs at about 1673 K, which is the durable temperature of the heat-resistant SiC fiber, it is of immense importance to select a material with the following properties as the water vapor shielding layer; significant wall-thinning resistance in the combustion gas environment, and coefficients of thermal expansion (CTEs) being close to that of the SiC/SiC. Klemm conducted burner-rig tests for various heat-resistant materials under the condition simulating the actual combustion gas environment to quantitatively evaluate relationship between CTEs and wall-thinning rate [8]. The result indicated that ytterbium silicate (Yb_2SiO_5; YbMS and $Yb_2Si_2O_7$; YbDS) has prominent wall-thinning resistance. He reported that the resistance of YbMS is particularly outstanding and that the CTE of YbMS is slightly larger than that of the SiC/SiC substrate, which is almost equal to that of YbDS. Yb-silicate is unlikely to fracture by phase transition during thermal cycle because phase transition causing the large volume change does not appear in the material throughout a wide temperature range from room temperature to high temperature. On the basis of these features, Yb-silicate is considered to be an attractive material as the water vapor shielding layer. By the way, YbMS is expected to be applied as the thermal barrier coating (TBC) for SiC/SiC substrate because the thermal conductivity of YbMS [9] is smaller than that of Y_2O_3-ZrO_2 (YSZ), which is used as TBC for Nickel-based super alloys.

Under the combustion gas environment, EBCs are exposed to a steep gradient of oxygen chemical potential ($d\mu_O$). It is known that, in an oxide layer under the environment of high temperature and $d\mu_O$, the oxide ion migrates from high oxygen partial pressure (P_{O_2}) to low P_{O_2} while the cation migrates in the opposite direction following the Gibbs-Duhem equation. This suggests that oxygen permeation occurs because of migration of cations as well as oxide ions. In addition, the cation migration may promote phase decomposition and fracture of the multi-layer component. For improving the phase stability of EBCs, therefore, it is of primary importance to prevent the migration of cations. Kitaoka et al. conducted oxygen permeability tests for representative heat-resistant oxides ($Al_6Si_2O_{13}$ (mullite) and Yb-silicate) under high temperature [10–12]. They clarified that the oxygen shielding of mullite was superior to that of Yb-silicate and that the mullite was decomposed due to the motion of Al ions from the low-P_{O_2} to high-P_{O_2} side producing deficiency of Al ions in the layer on the low-P_{O_2} side. Kitaoka et al. demonstrated that the phase of mullite is stabilized by depositing a SiAlON layer on the low-P_{O_2} side of the mullite layer because it feeds Al ions [13].

On the basis of the abovementioned studies, we propose an EBC with a function of thermal barrier as shown in Figure 1, which we call T/EBC hereafter. The structure consists of layers as follows (in order from the side of SiC/SiC substrate); the SiC layer for smoothing the surface of the substrate, the SiAlON layer for bonding with the substrate and ensuring structural stability of the mullite, the mullite dense layer for oxygen shielding, the Yb-silicate dense layer for water vapor shielding, and the layer of YbMS with porous segment structure for thermal shock resistance. Here, for decreasing thermal stress that occurs by the difference in CTEs of the coating layers and the substrate, the Yb-silicate dense layer is formed with gradient composition to have YbDS and YbMS on the substrate and top sides, respectively. Hereafter it is referred to as the Yb-silicate gradient composition layer. The topmost YbMS layer with a porous segment structure (topcoat) is employed for thermal shock resistance so as to relax strain due to rapid temperature change in the thickness direction associated with start and stop of the engine.

T/EBC is fabricated by depositing these layers on a SiC/SiC substrate at high temperature and cooling it to room temperature. During the cooling process, the thermal stress may cause interface crack initiation/propagation and delamination along interfaces between the layers with the function of environmental barrier. This leads to deterioration of the mechanical reliability of T/EBC. Thus, it is essential to design T/EBC in such a way that crack initiation and propagation are prevented during the cooling process.

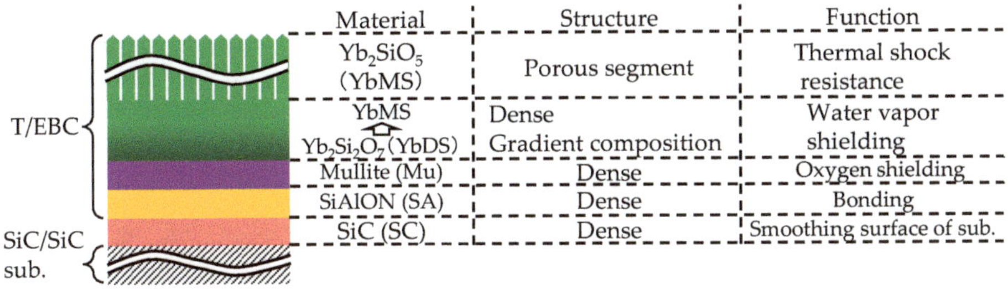

Figure 1. Schematic of thermal/environmental barrier coating (T/EBC) consisting of SiAlON (SA; bonding layer), mullite (Mu; oxygen shielding), layer of gradient composition of Yb_2SiO_5 (YbMS) and $Yb_2Si_2O_7$ (YbDS) (Yb-silicate gradient composition layer; water vapor shielding) and top YbMS with porous segment structure (thermal shock resistance).

In general, fracture of brittle materials follows the Griffith theory; i.e., a crack initiates or propagates when the energy release rate (ERR) exceeds fracture toughness for the corresponding fracture mode. While fracture toughness is a material property that should be evaluated in experiments, ERR can be calculated by numerical simulations through, for example, finite element method (FEM) calculations. Unlike conventional TBCs having simple structures, the mechanical behavior of T/EBC is expected to be complicated due to its complex structure consisting of multiple layers on a substrate. Numerical simulation can be an efficient tool to evaluate ERR even in such a complex structure.

In designing T/EBC structure with sufficient mechanical reliability, we must optimize a wide range of parameters, including conditions of multi-layer deposition processes and structural dimensions such as layer thicknesses. Layer thicknesses should affect much the mechanical state and play a crucial role in the mechanical stability of T/EBC under thermomechanical loading, which the component should undergo in operation. It is of importance to understand how such structural parameters influence the mechanical reliability. The aim of this study is to propose the condition of layer thicknesses for preventing interface crack initiation and propagation during the cooling process in the fabrication of the T/EBC. We perform FEM analysis for a T/EBC model with varying thicknesses of the coating layers to investigate the dependence of the layer thicknesses on ERR of crack initiation at interfaces, where thermal stress concentrates significantly. The analysis result facilitates the determination of layer thicknesses to prevent interface crack initiation. Then, the T/EBC is fabricated within the proposed thicknesses of the coating layers. We calculate ERR for interface crack initiation and propagation by conducting FEM analysis to a model of the fabricated T/EBC. Comparing the ERRs and fracture toughness obtained by interface fracture tests, we assess the validity of the proposed layer thicknesses.

2. Experiment and Simulation Procedures

2.1. Fabrication Procedure of T/EBC

In this study, the substrate was SiC/SiC (50 mm square and 3 mm in thickness) whose matrix was formed by chemical vapor infiltration and subsequent melt infiltration. Before T/EBC deposition, a SiC layer with an adequate thickness was deposited by chemical vapor deposition in order to smoothen the surface to be coated.

Each constituent layer of T/EBC shown in Figure 1 is a complex oxide or an oxynitride containing multiple cations. When these compounds are heated to a high temperature, the composition of vapor containing cation differs substantially from that of compounds in solid phase (incongruent vaporization). Thus, deposition methods accompanying the melting process of raw materials, such as plasma spraying, bring about the incongruent vaporization of the materials during deposition, resulting in significantly deviated compositions of coating layers.

In this study, therefore, dual electron beam-physical vapor deposition (EB-PVD) was employed to prepare layers of compounds which were evaporated incongruently. Two target raw materials located in the deposition chamber were evaporated using two electron guns regulated individually, which enabled independent control of vapor pressures of gases generated from the targets. Compound layers with arbitrary compositions were deposited on a substrate placed at a proper distance from the targets. During deposition, small amount of source gas was introduced in the coating chamber, and the coated surface was heated up to about 1473 K by irradiation of a direct diode laser (wavelength of 915 nm) to promote full crystallization and densification of each layer. After the deposition finished, the specimen was cooled from the deposition temperature to room temperature at the average cooling rate of 20 K/min.

Table 1 lists the target materials (Ingots A and B) and source gases used in the deposition of the layers. During deposition of the Yb-silicate gradient composition layer, the ratio of two electron beam powers irradiated to two targets was gradually changed with coating time to control the composition toward the thickness direction. The YbMS topcoat was deposited on a rotating sample for out-of-surface to form the porous segmented structure owing to the shadowing effect.

Table 1. Raw material targets and source gases for preparation of T/EBC.

Coating Layer	Raw Material Targets		Source Gas
	Ingot A	Ingot B	
SA	Al_2O_3	Si	NH_3
Mu	Al_2O_3	SiO_2	O_2
Yb-silicate gradient composition	Yb_2O_3	SiO_2	O_2
YbMS with porous segment structure	Yb_2O_3	SiO_2	O_2

The cross-sectional microstructure of T/EBCs was observed using a scanning electron microscope (SEM; SU8000 SEM, HITACHI, Tokyo, Japan) with energy dispersive spectroscopy (EDS).

2.2. ERR Calculation by FEM Analysis

For ensuring the mechanical reliability of T/EBC in the fabrication procedure, we need to determine its layer thicknesses with which ERR for interface crack initiation due to the thermal stress, G_{init}^{th}, can be reduced so that interface crack is not likely to initiate. In general, it is difficult to fabricate T/EBC without any initial interface crack such as microcrack. Thus, under the proposed layer thicknesses for preventing interface crack initiation, evaluating the behavior of interface crack propagation is also required. This means that we need to calculate ERR for interface crack propagation due to the thermal stress, G_{prop}^{th}, and compare with it and interface fracture toughness.

In this study, in order to investigate the effect of layer thicknesses on G_{init}^{th}, we conducted two-step FEM analyses using a T/EBC model with varying layer thickness; (1) Thermal stress analysis: FEM analysis for evaluating the thermal stress state after the cooling process and (2) Interface crack introduction analysis: FEM analysis for calculating ERR by introducing interface crack in the thermal stress state that was obtained in Step (1). Then, for obtaining G_{prop}^{th}, we conducted the two-step FEM analyses to the T/EBC model with the optimized layer thickness to prevent interface crack initiation. ABAQUS (ver.6.14.6, Dassault Systemes, Vélizy-Villacoublay, France) was used in the calculations.

2.2.1. FEM Model of T/EBC

The state of thermal stress in T/EBC is affected by the thicknesses of the coating layers. In particular, the difference in CTEs of the SiAlON and mullite layers is large compared to those at other interfaces (CTEs are explained in Section 2.2.2), and thus the thicknesses of the SiAlON and mullite layers are expected to significantly affect the state of thermal stress after cooling. We examined the simulation models mimicking T/EBC with various thicknesses of the SiAlON and mullite layers as shown in

Figure 2. Note that the FEM model used in this analysis is a model of the vicinity of interface edge of T/EBC deposited sample, which is indicated by a blue rectangle in Figure 2.

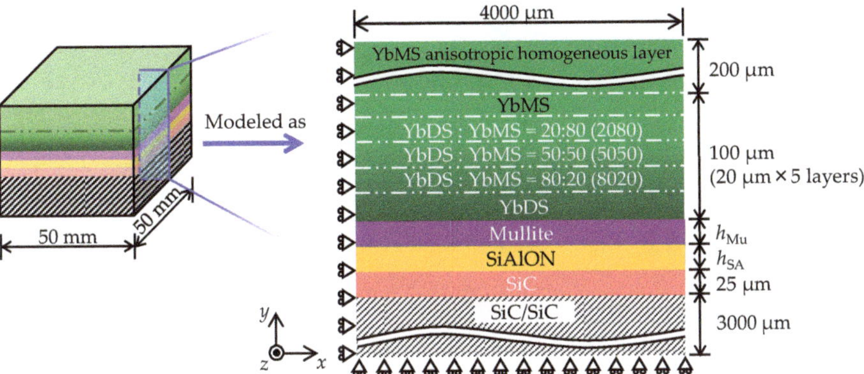

Figure 2. Schematic illustration of finite element method (FEM) analyses model and boundary condition.

In this study, the Yb-silicate gradient composition layer was modeled with five layers with different ratios of the YbDS and YbMS as follows; YbDS:YbMS = 100:0, 80:20, 50:50, 20:80, and 0:100. For computational efficiency and simplicity, the segmented YbMS layer was regarded as a homogenized material with an anisotropic (in-plane isotropic) mechanical property equivalent to the segmented layer [14]; hereafter it is referred to as the YbMS anisotropic homogeneous layer (Figure 2). The SiC/SiC substrate was treated as an in-plane isotropic material owing to the orientation of SiC fibers. T/EBC structure was represented by a 2-D simulation model, where displacements of the left and bottom edges of the model along the x and y directions, respectively, (shown in Figure 2) were constrained. A plane strain state was assumed to the z direction. In this analysis, the thicknesses of the SiAlON and mullite layers were changed (5, 15 and 25 μm), while the thicknesses of other layers were fixed as below; the SiC, Yb-silicate gradient composition and topcoat YbMS layers were 25, 100 (20 μm times five layers) and 200 μm, respectively. The thickness and width of the SiC/SiC substrate were 3000 and 4000 μm, respectively. The regions near all interfaces, where strong stress concentration is expected, were divided into a finer mesh (0.5 μm × 0.5 μm).

2.2.2. Thermal Stress Analysis

Since the experimental temperature is decreased gradually during the cooling process in fabrication of the T/EBC as explained in Section 2.1, the temperature dependence of the elastic properties (Young' modulus and Poisson's ratio) and the thermal property (CTE) must be taken into account for the thermal stress analysis. In addition, it is necessary to take account of the effect of creep on the mechanical state for the conditions of the experimental temperature and the cooling rate in the present deposition process. Thus, in this study, a series of the thermal stress analyses was carried out to evaluate the thermal stress distribution in T/EBC, where the creep effect was considered. Thermal stress is accumulated in T/EBC throughout the cooling process; i.e., stress is assumed to be null at the high temperature in the deposition process, and we evaluate the thermal stress state in T/EBC after cooling to room temperature from the high temperature. The thermal stress is dominated by stress in the x-axis direction in Figure 2 mainly, and thus the loading parallel to the interface, in other words, mode II-rich thermal loading is applied to T/EBC.

The temperature condition for this analysis was determined from the experimental condition of the cooling process after deposition of T/EBC (cooling from a high temperature in the deposition process to room temperature at cooling rate of 20 K/min). A uniform temperature distribution in T/EBC was assumed during the cooling process. Although the surface temperature of the coated layers is up to 1473 K in the deposition process as described in Section 2.1, in this analysis we set the deposition temperature (the initial temperature) to be 1673 K, which is an assumed operation temperature of T/EBC and provides an estimation on the safer side. In addition, the cooling rate of 20 K/min can cause a considerable effect of creep during the cooling process. In general, it is known that creep emerges significantly at temperatures higher than half the melting point [15]. Thus, in this analysis, the stress was set to be null at the initial state of 1673 K, and the development of stress distribution was calculated at a cooling rate of 20 K/min while creep was considered within the temperature range of 1673–1173 K. Creep was assumed to be negligible for SiC and SiC/SiC substrate. Since creep was not taken into account for all materials (i.e., the analysis is independent of time) below 1173 K, the system was cooled immediately from 1173 to 303 K.

As we consider a wide temperature range, we adopted temperature-dependent material properties; Young's modulus, E, Poisson's ratio, v, CTE (reference temperature: 1673 K), α', and creep properties (creep coefficient, C, and creep index, n) listed in Tables 2–8, respectively. The elastic properties (i.e., E and v) of the YbMS anisotropic homogeneous layer and the SiC/SiC substrate were regarded as anisotropic from the microstructure and the fiber orientation, respectively. Also, E and v of the mullite layer were regarded as anisotropic on the basis of the experimental result of the in-plane and out-of-plane loading tests on the deposited mullite layer (along the x and y axes in Figure 2, respectively). The other layers were dealt with as isotropic materials. CTE and the creep properties of each layer and substrate were assumed to be isotropic. The temperature-dependent material properties were calculated from experimental measurements and literature (see Appendix A).

Table 2. Young's moduli (E) of isotropic elastic layers.

Temp. T [K]	E [GPa]						
	SC	SA	YbDS:YbMS				
			100:0	80:20	50:50	20:80	0:100
303	476	232	128	121	111	102	96.0
373	474	231	128	121	111	102	95.9
473	472	229	127	120	110	101	95.8
573	469	227	126	119	110	101	95.5
673	467	226	125	118	109	100	95.0
773	464	224	124	117	108	100	94.5
873	462	223	123	116	107	99.0	93.8
973	459	221	122	115	106	98.1	93.0
1073	457	220	120	114	105	97.2	92.1
1173	454	218	119	113	104	96.2	91.2
1273	451	217	118	112	103	95.1	90.2
1373	449	215	116	110	102	94.0	89.2
1473	446	213	115	109	101	92.9	88.1
1573	444	212	113	108	99.3	91.7	86.9
1673	441	210	112	106	98.0	90.4	85.7

Table 3. Poisson's ratios (ν) of isotropic elastic layers.

Temp.			ν				
T [K]	SC	SA	YbDS:YbMS				
			100:0	80:20	50:50	20:80	0:100
303	0.21	0.34	0.26	0.25	0.25	0.24	0.23 [1]
373	0.21	0.35	0.25	0.25	0.24	0.23	–
473	0.21	0.35	0.25	0.25	0.24	0.23	–
573	0.21	0.35	0.25	0.25	0.24	0.23	–
673	0.21	0.35	0.25	0.25	0.24	0.23	–
773	0.21	0.35	0.25	0.25	0.24	0.23	–
873	0.21	0.36	0.25	0.25	0.24	0.23	–
973	0.21	0.39	0.25	0.25	0.24	0.23	–
1073	0.21	0.39	0.24	0.24	0.24	0.23	–
1173	0.21	0.39	0.24	0.24	0.24	0.23	–
1273	0.21	0.39	0.23	0.23	0.23	0.23	–
1373	0.21	0.39	0.23	0.23	0.23	0.23	–
1473	0.20	0.39	0.23	0.23	0.23	0.23	–
1573	0.19	0.39	0.23	0.23	0.23	0.23	–
1673	0.19	0.39	0.23	0.23	0.23	0.23	–

[1] No temperature dependence is assumed.

Table 4. Elastic properties (E, ν, and G) of SiC/SiC substrate.

Temp.	E [GPa]		ν			G [GPa]	
T [K]	E_x	E_y	ν_{xy}	ν_{xz}	ν_{yz}	G_{xy}	G_{xz}
303	150	59.9	0.25 [1]	0.1 [1]	0.08 [1]	37.5	68.1
373	147	58.9	–	–	–	36.8	67.0
473	144	57.5	–	–	–	36.0	65.4
573	140	56.1	–	–	–	35.1	63.8
673	137	54.7	–	–	–	34.2	62.2
773	133	53.3	–	–	–	33.3	60.6
873	130	51.9	–	–	–	32.4	59.0
973	126	50.5	–	–	–	31.5	57.4
1073	123	49.1	–	–	–	30.7	55.7
1173	119	47.7	–	–	–	29.8	54.1
1273	116	46.3	–	–	–	28.9	52.5
1373	112	44.8	–	–	–	28.0	50.9
1473	109	43.4	–	–	–	27.1	49.3
1573	105	42.0	–	–	–	26.3	47.7
1673	101	40.6	–	–	–	25.4	46.1

[1] No temperature dependence is assumed.

Table 5. Elastic properties (E, ν, and G) of mullite layer.

Temp.	E [GPa]		ν			G [GPa]	
T [K]	E_x	E_y	ν_{xy}	ν_{xz}	ν_{yz}	G_{xy}	G_{xz}
303	103	214	0.195 [1]	0.3 [1]	0.405 [1]	55.0	39.6
373	102	213	–	–	–	54.8	39.4
473	102	211	–	–	–	54.3	39.1
573	101	209	–	–	–	53.8	38.7
673	100	207	–	–	–	53.3	38.4
773	98.9	205	–	–	–	52.8	38.0
873	98.0	204	–	–	–	52.3	37.7

Table 5. *Cont.*

| Temp. | E [GPa] | | ν | | | G [GPa] | |
T [K]	E_x	E_y	ν_{xy}	ν_{xz}	ν_{yz}	G_{xy}	G_{xz}
973	97.0	202	–	–	–	51.8	37.3
1073	96.1	200	–	–	–	51.3	36.9
1173	95.1	198	–	–	–	50.8	36.6
1273	94.1	196	–	–	–	50.3	36.2
1373	93.2	194	–	–	–	49.8	35.8
1473	92.2	192	–	–	–	49.3	35.5
1573	91.2	190	–	–	–	48.8	35.1
1673	90.3	188	–	–	–	48.2	34.7

[1] No temperature dependence is assumed.

Table 6. Elastic properties (E, ν, and G) of YbMS anisotropic homogenous layer.

| Temp. | E [GPa] | | ν | | | G [GPa] | |
T [K]	E_x	E_y	ν_{xy}	ν_{xz}	ν_{yz}	G_{xy}	G_{xz}
303	0.135	96	0.003 [1]	0.24 [1]	0.23 [1]	0.135	0.055
373	0.135	95.9	–	–	–	0.135	0.054
473	0.135	95.8	–	–	–	0.135	0.054
573	0.135	95.5	–	–	–	0.134	0.054
673	0.134	95.0	–	–	–	0.134	0.054
773	0.133	94.5	–	–	–	0.133	0.054
873	0.132	93.8	–	–	–	0.132	0.053
973	0.131	93.0	–	–	–	0.131	0.053
1073	0.130	92.1	–	–	–	0.130	0.052
1173	0.129	91.2	–	–	–	0.128	0.052
1273	0.127	90.2	–	–	–	0.127	0.051
1373	0.126	89.2	–	–	–	0.125	0.051
1473	0.124	88.1	–	–	–	0.124	0.050
1573	0.122	86.9	–	–	–	0.122	0.049
1673	0.121	85.7	–	–	–	0.121	0.049

[1] No temperature dependence is assumed.

Table 7. Coefficients of thermal expansion (CTEs) at reference temperature of 1673 K (α') of substrate and coating layers.

| Temp. | α' [$\times 10^{-6}$/K] | | | | | | | | | |
| T [K] | Sub. | SC | SA | Mu | YbDS:YbMS | | | | | Ani. |
					100:0	80:20	50:50	20:80	0:100	
323	5.10	4.99	3.61	6.25	4.62	4.75	4.96	5.20	5.39	5.39
373	5.16	5.05	3.68	6.34	4.72	4.85	5.07	5.34	5.54	5.54
473	5.28	5.17	3.81	6.50	4.90	5.05	5.31	5.61	5.84	5.84
573	5.40	5.28	3.92	6.67	5.09	5.26	5.54	5.88	6.14	6.14
673	5.51	5.37	4.02	6.84	5.27	5.46	5.78	6.15	6.44	6.44
773	5.63	5.48	4.10	7.01	5.46	5.66	6.01	6.42	6.74	6.74
873	5.75	5.53	4.15	7.18	5.65	5.87	6.25	6.70	7.04	7.04
973	5.86	5.63	4.25	7.35	5.83	6.07	6.48	6.97	7.34	7.34
1073	5.98	5.73	4.35	7.52	6.02	6.28	6.72	7.24	7.64	7.64
1173	6.10	5.83	4.45	7.69	6.20	6.48	6.96	7.51	7.94	7.94
1273	6.21	5.93	4.55	7.86	6.39	6.69	7.19	7.78	8.24	8.24
1373	6.33	6.03	4.65	8.03	6.57	6.89	7.43	8.05	8.54	8.54
1473	6.45	6.13	4.75	8.20	6.76	7.09	7.66	8.33	8.84	8.84
1573	6.56	6.23	4.85	8.37	6.95	7.30	7.89	8.60	9.14	9.14
1693	6.70	6.35	4.97	8.57	7.17	7.54	8.18	8.92	9.50	9.50

Table 8. Creep properties (C and n) of coating layers.

Temp. T [K]	C^1							
	SA	Mu	YbDS:YbMS					Ani.
			100:0	80:20	50:50	20:80	0:100	
1173	2.62×10^{-15}	2.66×10^{-20}	2.26×10^{-12}	1.95×10^{-12}	1.57×10^{-12}	1.27×10^{-12}	1.10×10^{-12}	1.10×10^{-12}
1223	2.42×10^{-14}	9.45×10^{-19}	1.25×10^{-11}	1.10×10^{-11}	9.08×10^{-12}	7.50×10^{-12}	6.60×10^{-12}	6.60×10^{-12}
1273	1.88×10^{-13}	2.53×10^{-17}	6.05×10^{-11}	5.41×10^{-11}	4.57×10^{-11}	3.86×10^{-11}	3.44×10^{-11}	3.44×10^{-11}
1323	1.25×10^{-12}	5.29×10^{-16}	2.60×10^{-10}	2.35×10^{-10}	2.03×10^{-10}	1.75×10^{-10}	1.59×10^{-10}	1.59×10^{-10}
1373	7.24×10^{-12}	8.85×10^{-15}	1.00×10^{-9}	9.22×10^{-10}	8.10×10^{-10}	7.13×10^{-10}	6.54×10^{-10}	6.54×10^{-10}
1423	3.71×10^{-11}	1.22×10^{-13}	3.53×10^{-9}	3.28×10^{-9}	2.93×10^{-9}	2.63×10^{-9}	2.44×10^{-9}	2.44×10^{-9}
1473	1.70×10^{-10}	1.40×10^{-12}	1.14×10^{-8}	1.07×10^{-8}	9.73×10^{-9}	8.86×10^{-9}	8.32×10^{-9}	8.32×10^{-9}
1523	7.04×10^{-10}	1.37×10^{-11}	3.40×10^{-8}	3.23×10^{-8}	2.99×10^{-8}	2.76×10^{-8}	2.62×10^{-8}	2.62×10^{-8}
1573	2.67×10^{-9}	1.16×10^{-10}	9.48×10^{-8}	9.08×10^{-8}	8.53×10^{-8}	8.00×10^{-8}	7.67×10^{-8}	7.67×10^{-8}
1623	9.31×10^{-9}	8.61×10^{-10}	2.48×10^{-7}	2.40×10^{-7}	2.28×10^{-7}	2.17×10^{-7}	2.10×10^{-7}	2.10×10^{-7}
1673	3.01×10^{-8}	5.67×10^{-9}	6.12×10^{-7}	5.98×10^{-7}	5.76×10^{-7}	5.55×10^{-7}	5.42×10^{-7}	5.42×10^{-7}
n^2								
–	SA	Mu	YbDS:YbMS					Ani.
			100:0	80:20	50:50	20:80	0:100	
	1.09	1.20	0.960	0.976	1.00	1.02	1.04	1.04

^{1} The unit of C depends on the corresponding creep index n. ^{2} No temperature dependence is assumed.

2.2.3. Interface Crack Introduction Analysis

ERR owing to the thermal stress, $\mathcal{G}^{\text{th}}$, is defined as the reduction in strain energy, Π, stored in a structure per increased crack area. When the crack area, A, increases to $A + \mathrm{d}A$, ERR is evaluated as follows:

$$\mathcal{G}^{\text{th}} = -\frac{\mathrm{d}\Pi}{\mathrm{d}A}. \tag{1}$$

Thus, it is necessary to obtain the strain energies stored in T/EBC with different crack lengths to evaluate $\mathcal{G}^{\text{th}}_{\text{init}}\left(= \mathcal{G}^{\text{th}}\big|_{A \to 0}\right)$ and $\mathcal{G}^{\text{th}}_{\text{prop}}$. We carried out FEM analyses for T/EBC models with various crack length under the stress condition obtained from the thermal stress analysis, and the strain energy in T/EBC was evaluated as a function of crack length.

The geometry, the boundary conditions and the mesh sizes of the model in this analysis were set in the same way as in the thermal stress analysis. The initial stress condition was derived from the stress distribution of the uncracked T/EBC model obtained by the thermal stress analysis. The temperature was kept constant at 303 K (temperature after the cooling process) during the analyses, and thus we used the material properties E and v at 303 K shown in Tables 2–6. Interface cracks were introduced from the right edge of the simulation model shown in Figure 2. According to a preliminary FEM analysis, the SiC/SiAlON, SiAlON/mullite and mullite/YbDS interfaces were chosen as the objective interfaces to evaluate $\mathcal{G}^{\text{th}}_{\text{init}}$ and $\mathcal{G}^{\text{th}}_{\text{prop}}$ (the detail of the preliminary analysis is found in Appendix B).

2.2.4. Evaluation of ERR for Interface Crack Initiation and Propagation

We evaluated $\mathcal{G}^{\text{th}}_{\text{init}}$ and $\mathcal{G}^{\text{th}}_{\text{prop}}$ from the strain energy in T/EBC with different crack lengths obtained by the FEM analyses in Section 2.2.3 in the following ways.

Evaluation of $\mathcal{G}^{\text{th}}_{\text{init}}$

$\mathcal{G}^{\text{th}}_{\text{init}}$ is defined as ERR in the limit when A approaches 0, as follows:

$$\mathcal{G}^{\text{th}}_{\text{init}} = -\lim_{A \to 0} \frac{\mathrm{d}\Pi}{\mathrm{d}A}. \tag{2}$$

Note that $\mathcal{G}^{\text{th}}_{\text{init}}$ is equal to the first-order coefficient of the Taylor expansion of $\Pi(A)$ around $A = 0$ with the inverted sign. Therefore, we approximated $\Pi(A)$ as a polynomial from the FEM analyses for the T/EBC models with several different crack lengths, and $\mathcal{G}^{\text{th}}_{\text{init}}$ was evaluated as the first-order coefficient of the polynomial with the inverted sign. Two cases are possible according to the sign of $\mathcal{G}^{th}_{init}$: for $\mathcal{G}^{th}_{init} > 0$, crack initiation occurs if the interface fracture toughness, Γ, is smaller than $\mathcal{G}^{th}_{init}$; for $\mathcal{G}^{th}_{init} < 0$, no crack initiation is expected. From the result of the preliminary analysis (Appendix B), we examined crack length a = 2.5, 5.0 and 7.5 µm to evaluate $\mathcal{G}^{th}_{init}$ at the objective interfaces.

Evaluation of $\mathcal{G}^{th}_{prop}$

ERR when a crack propagates from A to $A + \Delta A$ is given by numerical differentiation as follows:

$$\mathcal{G}^{th}_{prop}\left(A + \frac{\Delta A}{2}\right) = -\left.\frac{d\Pi}{dA}\right|_{A+\Delta A/2} \approx -\frac{\Pi(A + \Delta A) - \Pi(A)}{\Delta A}. \tag{3}$$

From the result of the preliminary analysis (Appendix B), the a for evaluation of $\mathcal{G}^{th}_{prop}$ was determined as 2.5, 5.0, 7.5, 10, 50, 100, 250, ..., 3500 µm in increments of 250 µm between 250 and 3500 µm.

2.3. Interface Fracture Test

Interfacial toughness was measured by an interface fracture test. The test method was a modification of the one designed for EBCs on SiC/SiC composites having a weak inter-laminar strength [16]. Figure 3 shows schematic of the test set-up.

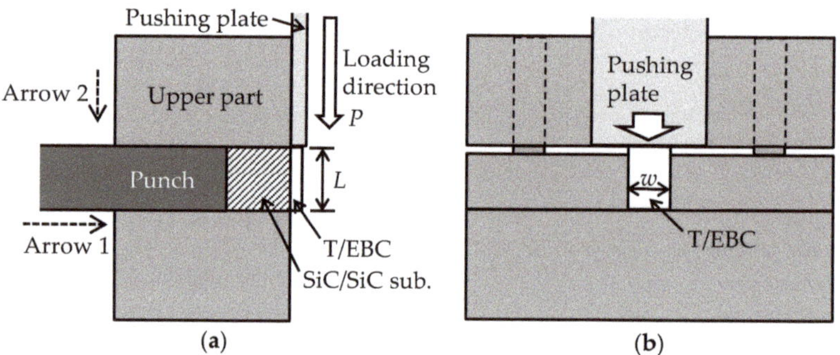

Figure 3. Schematics of experimental set-up of interface fracture test: (**a**) Cross-sectional side view; (**b**) Front view.

A block of specimen with a height, L, of 4 mm and a width, w, of 3 mm was prepared using a diamond blade saw in such a way that one of the fiber directions in the woven fabric of the SiC/SiC substrate was oriented to the longitudinal side of the specimen. The surface was as cut without polishing to avoid the coating delamination during the processing. The cut surface was smooth enough for us to distinguish the interface in the cross-sectional observation by conventional optical microscopy.

The specimen was set in a steel jig like a horizontal die, the upper part of which was adjustable. The specimen was put in the die and slid horizontally from the back side by a punch as indicated by the dashed arrow 1 in Figure 3 so that only the coating protruded from the die while the substrate was buried. Then the upper part was adjusted in the vertical direction (the dashed arrow 2 in Figure 3) to fix the specimen. The jig with the fixed specimen was placed under a universal testing machine (load cell capacity: 500 N, EZ-LX, Shimadzu Corporation, Kyoto, Japan) having a steel pushing plate with a thickness of 0.5 mm and a width of 5 mm.

The edge of the coating was compressively loaded by the plate with a constant crosshead speed of 0.024 mm/min till the coating was delaminated from the substrate; thus, the mode II-rich loading at the interface was achieved with little loading to the substrate in the inter-laminar shear direction. The shear stress ought not to concentrate at only an interface between SiC/SiC substrate and T/EBC

(SiC/SiAlON interface); i.e., it occurs along some interfaces in T/EBC. The pushing load, P, and the crosshead displacement, u, were measured during the test. The coating near the loading point was observed in the cross-sectional direction of the specimen by optical microscopy to determine when a crack was initiated at the loaded edge and the corresponding load, P_{init}. The number of the tests was six.

The energy release rate, $\mathcal{G}_{\text{init}}$, associated with the initiation of an interface crack at the loaded edge under an applied load of P_{init} is given by

$$\mathcal{G}_{\text{init}} = \frac{(P_{\text{init}})^2}{2w^2 h_{\text{coat}}} \left\{ \frac{1}{E'_{\text{coat}}} - \frac{1}{E'_{\text{sub}}} \left(\frac{1}{D} + \frac{(h_{\text{coat}}/2 + h_{\text{sub}} - \delta)^2}{I(h_{\text{coat}})^2} \right) \right\}. \tag{4}$$

Definition of parameters and details of deriving Equation (4) are shown in Appendix C. The $\mathcal{G}_{\text{init}}$ is defined as fracture toughness for the interface crack initiation, Γ_{init}. Note that, Γ_{init} is a nominal fracture toughness value of interface crack initiation including the effect of thermal residual stress.

After the tests, four of the specimens were mounted in resin and their cross-sections of the delaminated coating and substrate were observed; the cross-section was polished to mirror finish. SEM observation (VE–7800, Keyence corporation, Osaka, Japan) and energy dispersive X-ray (EDX) analysis (EX74120 attached to field emission SEM (JSM6500F, JEOL Ltd., Tokyo, Japan)) of the cross section were done to identify the fractured interface. The remaining two specimens were served for the analyses of fracture surfaces; optical and SEM observations and EDX analysis were done.

3. Results of T/EBC Fabrication

3.1. Layer Thickness Condition for Preventing the Objective Interface Crack Initiation

Figure 4a–c show the relationships between the SiAlON layer thickness, h_{SA}, and $\mathcal{G}^{\text{th}}_{\text{init}}$ at the SiC/SiAlON, SiAlON/mullite and mullite/YbDS interfaces, respectively, with various mullite layer thicknesses, h_{Mu}. From Figure 4a, the ERR for interface crack initiation at the SiC/SiAlON interface, $\mathcal{G}^{\text{th}}_{\text{init, SC/SA}}$, is nearly zero for any combination of layer thicknesses. Thus, a crack is unlikely to initiate at the SiC/SiAlON interface during the cooling process after deposition if thicknesses of the SiAlON and mullite layers are both within the range of 5–25 μm.

The ERR for interface crack initiation at the SiAlON/mullite interface, $\mathcal{G}^{\text{th}}_{\text{init, SA/Mu}}$, depends on the thicknesses of h_{SA} and h_{Mu} as shown in Figure 4b, which is found to be expressed as the following equation within the examined range of h_{SA} and h_{Mu} (5–25 μm):

$$\mathcal{G}^{\text{th}}_{\text{init,SA/Mu}}(h_{\text{SA}}, h_{\text{Mu}}) = \sum_{k=0}^{2} \sum_{l=0}^{2} c_{kl} (h_{\text{SA}})^k (h_{\text{Mu}})^l \tag{5}$$

with the coefficients, c_{kl}, listed in Table 9. Figure 5 shows the contour chart of $\mathcal{G}^{\text{th}}_{\text{init, SA/Mu}}$ as a function of h_{SA} and h_{Mu} shown in Equation (5). The level of $\mathcal{G}^{\text{th}}_{\text{init, SA/Mu}}$ is indicated by the depth of color (violet); a deeper-colored region corresponds to a lower $\mathcal{G}^{\text{th}}_{\text{init, SA/Mu}}$ and thus a better combination of layer thicknesses. This result suggests that one of the SiAlON or mullite layers must be thick and the other be thin in order to decrease $\mathcal{G}^{\text{th}}_{\text{init, SA/Mu}}$ and hinder crack initiation at the SiAlON/mullite interface.

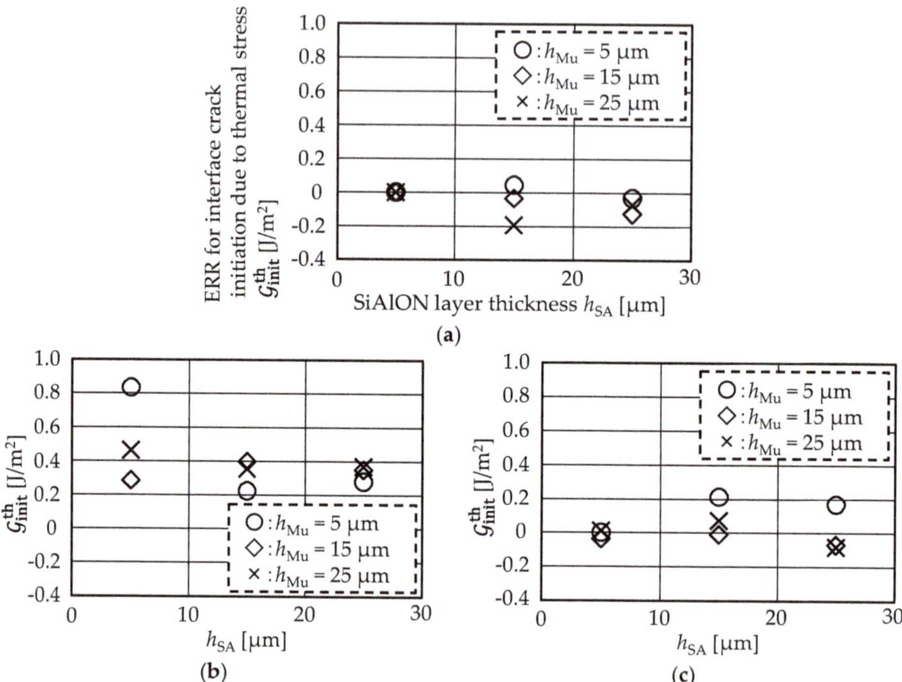

Figure 4. Relationship between SiAlON layer thickness and ERR for interface crack initiation due to thermal stress at (**a**) SiC/SiAlON, (**b**) SiAlON/mullite and (**c**) mullite/YbDS interfaces with various mullite layer thicknesses.

Table 9. Coefficients for polynomial approximation of $G^{\mathrm{th}}_{\mathrm{init,SA/Mu}}(h_{\mathrm{SA}}, h_{\mathrm{Mu}})$ in Equation (5), c_{kl} [1].

c_{kl}		l		
		0	**1**	**2**
	0	2.61	-2.83×10^{-1}	8.06×10^{-3}
k	**1**	-2.82×10^{-1}	3.60×10^{-2}	-1.03×10^{-3}
	2	7.48×10^{-3}	-9.68×10^{-4}	2.77×10^{-5}

[1] Unit of c_{kl}: $\mu m^{(k+l)}$.

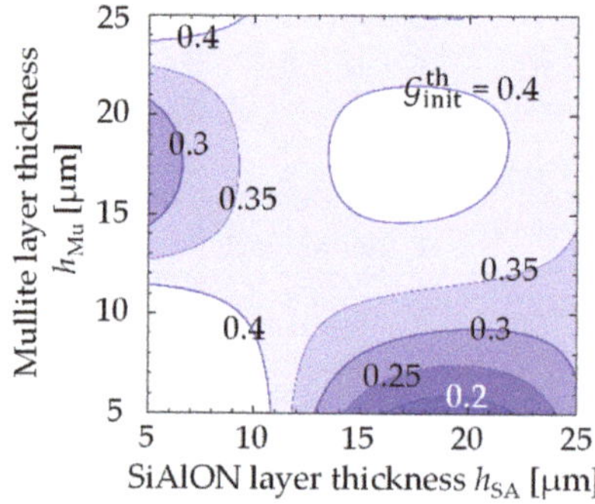

Figure 5. Contour chart of $G^{\mathrm{th}}_{\mathrm{init,SA/Mu}}(h_{\mathrm{SA}}, h_{\mathrm{Mu}})$ interpolated with Equation (5).

As shown in Figure 4c, it is found that a thin mullite layer (h_{Mu} = 5 µm) results in a rise in the ERR for interface crack initiation at the mullite/YbDS interface, $\mathcal{G}^{th}_{init,\,Mu/YD}$, with increasing h_{SA}, while a thicker mullite layer (h_{Mu} = 15 and 25 µm) reduces $\mathcal{G}^{th}_{init,\,Mu/YD}$ to almost zero regardless of h_{SA}. Therefore, a thick mullite layer is necessary to decrease $\mathcal{G}^{th}_{init,\,Mu/YD}$, i.e., to suppress crack initiation at the mullite/YbDS interface.

From the above results, it is found that a thin SiAlON layer and a thick mullite layer are required for reducing $\mathcal{G}^{th}_{init}$ and preventing crack initiation at the SiC/SiAlON, SiAlON/mullite and mullite/YbDS interfaces, where strong stress concentration is expected during the cooling process in fabrication of T/EBC as shown in Figure 1.

3.2. Phase Structure of T/EBC

In Section 3.1, we found that thinning the SiAlON layer together with thickening the mullite layer was effective in suppressing interface crack formation during cooling where a large thermal stress is expected, i.e., the SiC/SiAlON, SiAlON/mullite and mullite/YbDS interfaces. In this study, therefore, the thickness of the SiAlON layer and the mullite layer were determined to be 5 and 20 µm, respectively, and T/EBC was prepared by the dual EB-PVD process described in Section 2.1. We confirmed no delamination at all the interfaces of the fabricated T/EBC. Cross-sectional SEM images and elemental mapping images of T/EBC were shown in Figure 6a,b, respectively, where we found the thicknesses of constituent layers to be; SiAlON: about 5 µm, mullite: about 20 µm, Yb-silicate gradient composition layer: about 100 µm, and porous segmented YbMS layer: about 200 µm. That is, T/EBC was successfully fabricated nearly as designed.

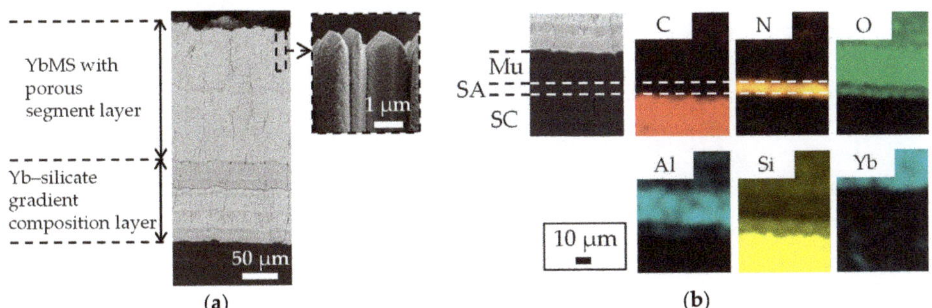

Figure 6. (**a**) Cross-sectional SEM images of T/EBC and (**b**) elemental mapping images focused on SiC/SiAlON/mullite layers.

4. Behavior of Interface Crack Initiation and Propagation in T/EBC

To investigate the behavior of crack initiation and propagation during the cooling process in fabrication of T/EBC shown in Section 3.2, calculated ERRs ($\mathcal{G}^{th}_{init}$, $\mathcal{G}^{th}_{prop}$) must be compared with interface fracture toughness obtained by experiment. Thus, we carried out the interface fracture toughness tests (Section 2.3) to obtain Γ, which is to be compared with $\mathcal{G}^{th}_{init}$ and $\mathcal{G}^{th}_{prop}$ with h_{SA} = 5 µm and h_{Mu} = 20 µm.

4.1. $\mathcal{G}^{th}_{init}$ and $\mathcal{G}^{th}_{prop}$ for Objective Interface Cracks in Fabricated T/EBC

Table 10 lists $\mathcal{G}^{th}_{init}$ of the objective interface cracks for the T/EBC model with h_{SA} = 5 µm and h_{Mu} = 20 µm. From this result, the SiAlON/mullite interface is found to possess the highest $\mathcal{G}^{th}_{init}$ of the objective interfaces.

Table 10. $\mathcal{G}^{\text{th}}_{\text{init}}$ for T/EBC model with $h_{\text{SA}} = 5$ µm and $h_{\text{Mu}} = 20$ µm.

Interface	$\mathcal{G}^{\text{th}}_{\text{init}}$ [J/m^2]
SiC/SiAlON	0
SiAlON/mullite	0.324
Mullite/YbDS	−0.069

Figure 7 shows the relationships between $\mathcal{G}^{\text{th}}_{\text{prop}}$ and a at the objective interfaces in the T/EBC with $h_{\text{SA}} = 5$ µm and $h_{\text{Mu}} = 20$ µm. In the $\mathcal{G}^{\text{th}}_{\text{prop}}$–$a$ relationships at all the objective interfaces, we observe three stages with increasing a: (A) $\mathcal{G}^{\text{th}}_{\text{prop}}$ shows a relatively rapid increase; (B) $\mathcal{G}^{\text{th}}_{\text{prop}}$ decreases; (C) $\mathcal{G}^{\text{th}}_{\text{prop}}$ increases again.

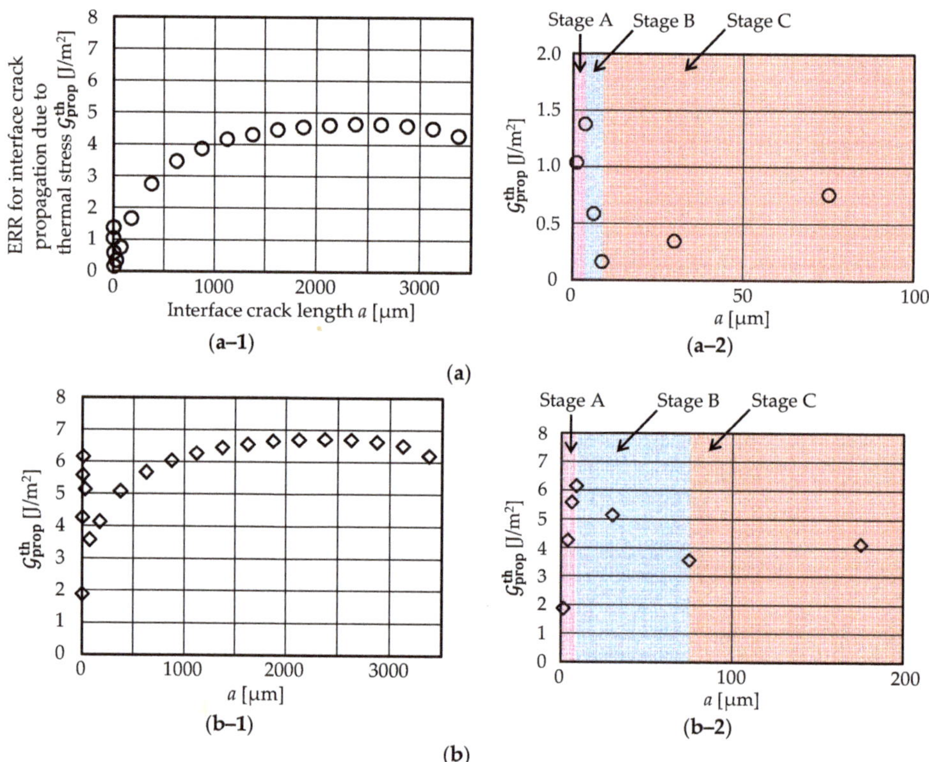

Figure 7. *Cont.*

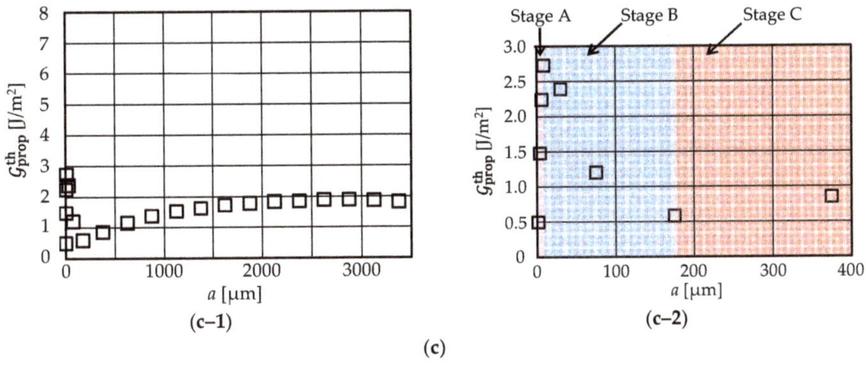

Figure 7. Relationships between ERR for interface crack propagation due to the thermal stress and crack length at objective interfaces in T/EBC with $h_{SA} = 5$ µm and $h_{Mu} = 20$ µm; (**a**) SiC/SiAlON interface: (**a–1**) $0 \leq a \leq 3500$ µm and (**a–2**) $0 \leq a \leq 100$ µm, (**b**) SiAlON/mullite interface: (**b–1**) $0 \leq a \leq 3500$ µm and (**b–2**) $0 \leq a \leq 200$ µm and (**c**) mullite/YbDS interface: (**c–1**) $0 \leq a \leq 3500$ µm and (**c–2**) $0 \leq a \leq 100$ µm.

The reason for the decrease in $\mathcal{G}_{prop}^{th}$ at Stage B is explained by the distribution of out-of-plane thermal stress in the vicinity of the interface edge. Figure 8a shows the σ_y distributions in the coating layers along the y axis, which are obtained at the positions r_x distant from the right interface edge in the x direction ($r_x = 0, 5, 10, 20, 30$ and 40 µm). The highest level of σ_y was observed at the interface edge ($r_x = 0$ µm), and the σ_y distribution reduces immediately to a negligible level within $r_x = 30$ µm. Thus, the effect of σ_y on the coating layers is significant but extremely localized in the vicinity of interface edge, while that is negligible inside the simulation model. Note that the effect of σ_x on $\mathcal{G}_{prop}^{th}$ is marginal for sufficiently small a because we find $\sigma_x \sim 0$ at the interface edge ($r_x = 0$ µm) directly from the balance of stress component in the x direction. Figure 8b shows the σ_x distributions in the coating layers along the y axis at various r_x. These results indicate that σ_y has a dominating effect on $\mathcal{G}_{prop}^{th}$ for very short cracks. The mechanism of the second rise in $\mathcal{G}_{prop}^{th}$ at Stage C is attributed to the increase in the level of σ_x inside the simulation model as shown in Figure 8b. Therefore, the three-stage behavior in the $\mathcal{G}_{prop}^{th}$–a relationship is explained as follows: in the beginning, $\mathcal{G}_{prop}^{th}$ rises as a crack propagates owing to the strong σ_y near the interface edge (Stage A); then $\mathcal{G}_{prop}^{th}$ decreases because of a steep drop in σ_y inside the model (Stage B); for a sufficiently long crack, $\mathcal{G}_{prop}^{th}$ is mainly affected by σ_x and rises again with increasing σ_x (Stage C).

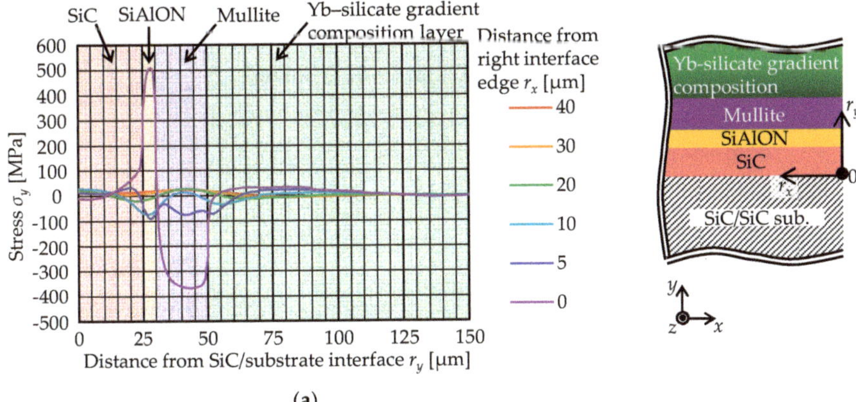

(**a**)

Figure 8. *Cont.*

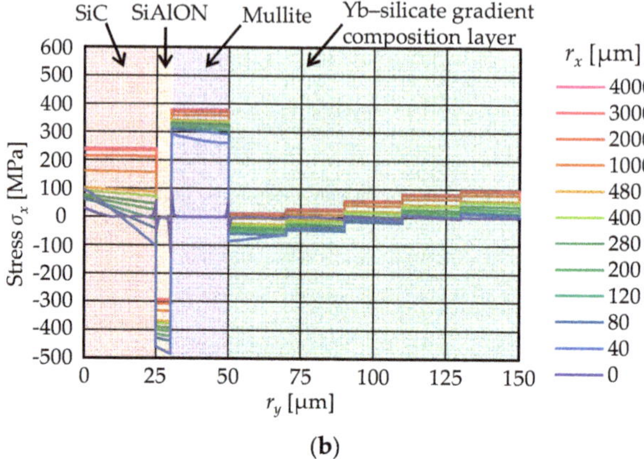

(b)

Figure 8. (a) σ_y distributions, and (b) σ_x distributions in coating layers along the y axis at the positions r_x distant from right SiC/substrate interface edge in the x direction.

4.2. Interface Fracture Toughness of T/EBC

Figure 9 shows a typical P–u curve during the tests. By the cross-sectional optical microscopy, the onset of interface fracture near the loaded edge was identified at Point A in P–u curve before the maximum load. The load at Point A is defined as P_{init}. The crack then propagated down through the interface to the midway till the load reached the maximum; finally it grew unstably to the complete delamination at the maximum load. Buckling or compressive fracture of the coating was not observed during the test. The measured P_{init} values for respective tests are listed in Table 11.

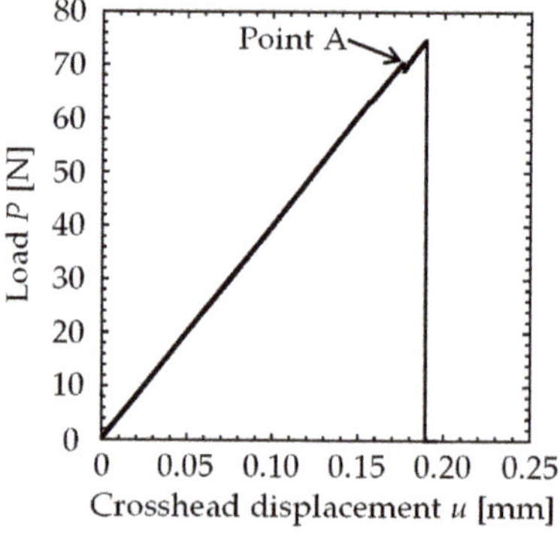

Figure 9. Typical load-crosshead displacement curve during test (Specimen No. 2).

Table 11. Crack initiation load (P_{init}) obtained by interface fracture test.

Specimen Number	P_{init} [N]
1	83
2	70
3	71
4	64
5	76
6	76
Average	73

Examples of the cross-sectional SEM images of the specimens after the tests were shown in Figure 10. At the loaded edge where the interface crack started to propagate, fracture occurred below or above the SiAlON layer, i.e., either at the interface between the SiC and SiAlON layers (Figure 10a) or at the interface between the SiAlON and mullite layers (Figure 10b). The crack path was along either of the two interfaces, and it was switched to each other by crack kinking across the SiAlON layer. However, crack propagation within a layer, such as a crack passing through the inside of layer parallel to the coating plane, was not observed.

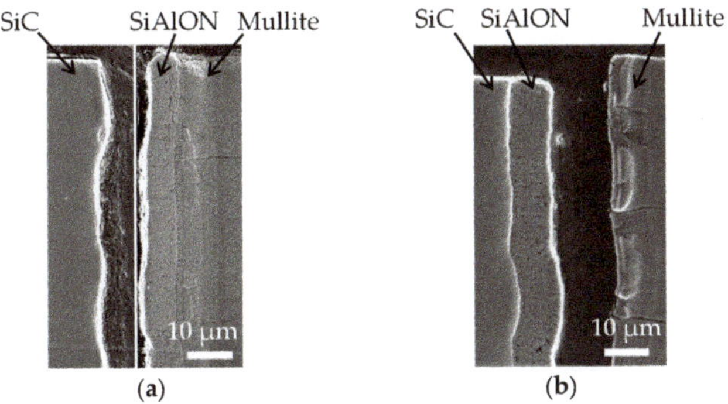

Figure 10. Examples of cross-sectional SEM images of specimens after interface fracture test; Fracture at (**a**) SiC/SiAlON interface and (**b**) SiAlON/mullite interface.

Figure 11a shows a fracture surface near the loaded edge after the tests observed by optical microscopy. The fracture surface was classified into three according to their colors: the largest black area denoted by Region A, the gray area by Region B, and the small white area by Region C as illustrated in Figure 11a. EDX maps of a selected area in the fracture surface (indicated by a dashed rectangle in Figure 11a) are shown in Figure 11b. Mapping of element distributions shows the concentration of Si in Region A. In Region B the existence of Si, Al and N was detected. Region C was characterized by Yb concentration. Apparently Al also looks rich in Region C, but it is due to misdetection because the M_α line of Yb is very close to K_α line of Al which was used for the mapping of Al distribution. These results suggest that Region A with the largest area corresponds to the fracture surface created by the interface fracture between the SiC and SiAlON layers observed in Figure 10a; Region B showing the second largest area corresponds to the surface created by the SiAlON/mullite interface fracture (Figure 10b). We could not observe the fractured interface corresponding to Region C in the cross-sectional SEM observation of the T/EBC and substrate (Figure 10) because of the limited area of Region C. To determine the exact location of the fracture surface in Region C, we cut after the tests one of the substrate in the cross-section including Region C. The result is shown in Figure 12 suggesting that the fracture occurred in the Yb-silicate gradient composition layer.

To calculate the line fractions of Regions A, B and C along the loaded edge, the fracture surfaces near the edge were analyzed using image processing software (Image J 1.48v, National Institute of Health, Bethesda, MD, USA) (Table 12). As shown in Table 12, fracture surface was dominantly composed of Regions A and B; the ratio of Region C was relatively small. Thus the crack initiation at the edge was supposed to occur primarily either at the interface between the SiC and SiAlON layers or at the interface between the SiAlON and mullite layers.

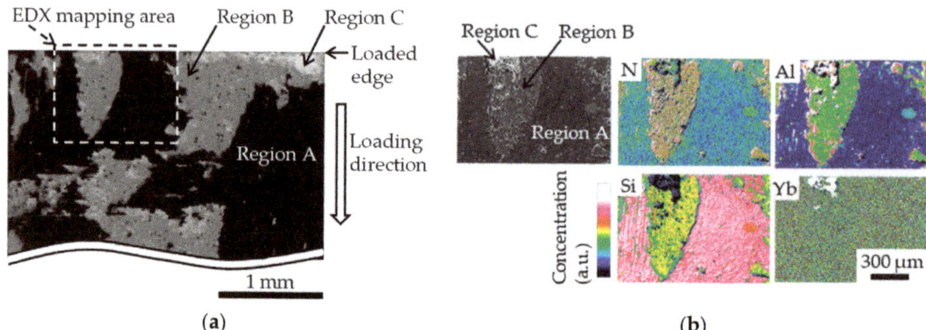

Figure 11. (**a**) Top view of fracture surface of specimen after interface fracture test and (**b**) EDX maps of selected area in fracture surface indicated by a dashed rectangle in (**a**).

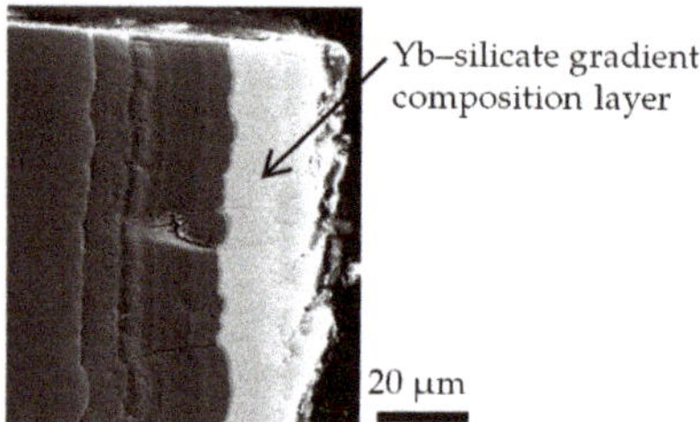

Figure 12. Cross-sectional SEM observation including Region C.

Table 12. Line fractions of Regions A, B and C at loaded edge.

Region	Line Fraction
A	0.31
B	0.56
C	0.13

The energy release rate for the edge crack initiation along the SiC/SiAlON interface under P_{init} (in Table 11) was calculated from Equations (A14)–(A18) and (4), regarding the SiC/SiC plate and the SiC layer as "substrate" and the rest of the layers as "coating" in Equation (A14); the mean value of 6.4 J/m^2 was obtained. When the SiAlON layer is counted as part of the substrate in addition to the SiC/SiC and SiC, the mean energy release rate for the crack along the SiAlON/mullite interface becomes 6.5 J/m^2. These values of energy release rates are very close to each other because the thickness of the SiAlON layer lying between the two interfaces is quite small compared to the total thickness of the system. Under the assumption that the edge crack was initiated simultaneously through the edge (i.e., in the direction of depth) at P_{init} regardless of the fractured interfaces, we can estimate that the SiC/SiAlON interface fracture and the SiAlON/mullite interface fracture both occur with significant proportions at almost the same energy release rates. This suggests that the interface toughness of these two interfaces are very close; both the interfaces should have an interface toughness of ~6.4 J/m^2. The mean energy release rate at P_{init} when the edge crack was initiated along the mullite/YbDS interface, which was supposed to have large thermal stress according to the FEM analysis, was calculated to be 6.8 J/m^2. Since no fracture occurred at this interface at P_{init}, we can expect that the toughness of the interface should be larger than 6.8 J/m^2. The abovementioned results indicate that the approximate Γ_{init} for the SiC/SiAlON and SiAlON/mullite interfaces are 6.4 J/m^2 and the lower limit of Γ_{init} for the mullite/YbDS interface is 6.8 J/m^2.

Table 13 shows comparison between $\mathcal{G}^{th}_{init}$ of the SiC/SiAlON, SiAlON/mullite and mullite/YbDS interface cracks and Γ_{init} of the corresponding interfaces. $\mathcal{G}^{th}_{init}$s of every objective interface crack are smaller than Γ_{init}. This result suggests that the effect of thermal residual stress on the intrinsic fracture toughness for interface crack initiation is sufficiently small. Therefore, in this study, we regard the nominal value as the fracture toughness of interface crack initiation. Table 13 also suggests that the SiC/SiAlON, SiAlON/mullite and mullite/YbDS interface cracks are not likely to initiate by cooling in the T/EBC fabrication process.

Table 13. Comparison between $\mathcal{G}^{th}_{init}$ of target interface cracks and fracture toughness for interface crack initiation (Γ_{init}) of corresponding interfaces.

Interface	$\mathcal{G}^{th}_{init}$ [J/m^2]	Γ_{init} [J/m^2]
SiC/SiAlON	0	6.4 [1]
SiAlON/mullite	0.324	6.4 [1]
Mullite/YbDS	−0.069	6.8 [2]

[1] Approximate value. [2] Lower limit value.

Figure 13a–c show comparisons between $\mathcal{G}^{th}_{prop}$ and fracture toughness for interface crack propagation, Γ_{prop}, at the SiC/SiAlON, SiAlON/mullite and mullite/YbDS interfaces. Here, we assumed that fracture surfaces formed by interface crack initiation and propagation were the same and regarded Γ_{prop} as equal to Γ_{init}. As shown in Figure 13a,c, $\mathcal{G}^{th}_{prop}$ of the SiC/SiAlON and mullite/YbDS interface cracks are smaller than Γ_{prop} of the corresponding interfaces. The results suggest that these interface cracks are not likely to propagate regardless of initial interface crack length. Figure 13b suggests that, if the T/EBC has an initial length of the SiAlON/mullite interface crack of about 1200 μm, the crack should propagate instantly because $\mathcal{G}^{th}_{prop}$ exceeds Γ_{prop} at the crack length of 1200 μm. However, delamination along the SiAlON/mullite interface was not observed in the experiment (see Section 3.2). These results suggest that there was no initial crack exceeding the length threshold at the SiAlON/mullite interface.

The results shown in this section reveal that the T/EBC with the proposed layer thicknesses can be fabricated without delamination along interfaces by cooling in the fabrication process. This is confirmed by the result that no delamination along interface is observed as described in Section 3.2.

4.3. Future Prospects

In this study, we mainly focused on the mechanical reliability of T/EBC during the cooling process in fabrication. From the practical viewpoint, however, it is also essential to evaluate the mechanical reliability under an operating condition. In general, an EBC is exposed to a thermal cycle condition with a high humidity in operation, which can induce a microstructural change and chemical transformation of the coating layers through a reaction with heated oxygen and water vapor. Consequently, changes in the material properties of each layer due to microstructural change and chemical transformation are crucial to the mechanical state in T/EBC. Thus, it is necessary to evaluate a time dependent ERR by the FEM analysis where changes in material properties over time are incorporated. Furthermore, the interface fracture toughness changes presumably over time owing to microstructural change and chemical transformation, which should also be taken into account for a reliable design of T/EBC. Nevertheless, it is beyond the scope of this paper and will be considered in our future work.

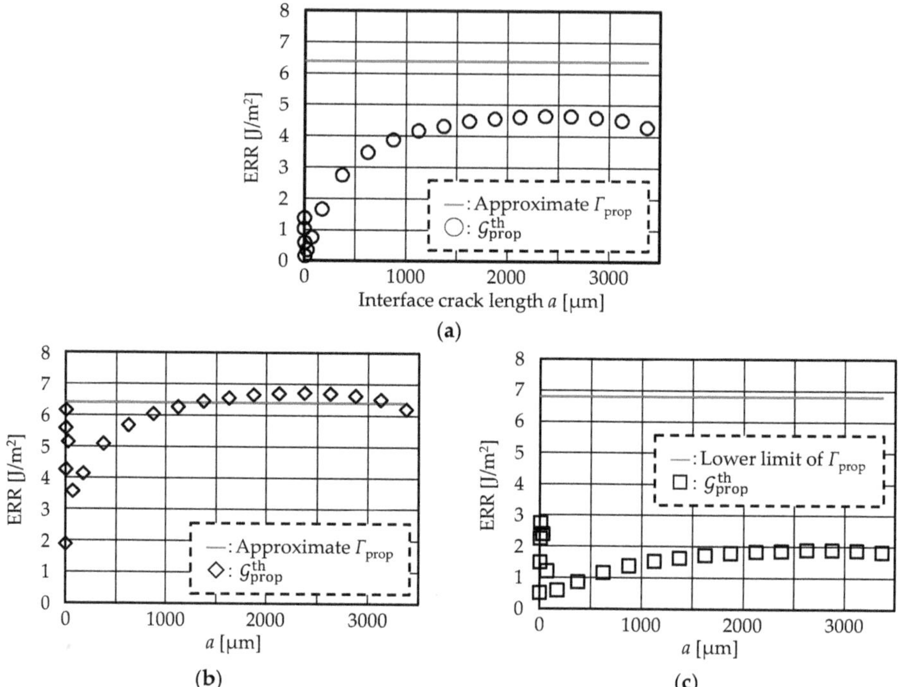

Figure 13. Comparison between ERR for crack propagation due to thermal stress and fracture toughness for crack propagation at (**a**) SiC/SiAlON interface, (**b**) SiAlON/mullite interface and (**c**) mullite/YbDS interface.

5. Conclusions

To assess the mechanical reliability of T/EBC (SiC/SiAlON/mullite/Yb-silicate gradient composition layer/YbMS with porous segment structure) against interface crack initiation and propagation due to thermal stress induced in fabricated process, we carried out FEM analysis to evaluate ERR for interface cracks and performed experiment to obtain interface fracture toughness. Our FEM calculations revealed that $\mathcal{G}_{init}^{th}$ of the objective interfaces decreases by making the SiAlON layer thinner and the mullite layer thicker. We fabricated the T/EBC with layer thicknesses within the proposed range ($h_{SA} = 5$ µm and $h_{Mu} = 20$ µm) and confirmed no delamination along the interfaces. In the interface fracture test for the fabricated T/EBC, fracture surfaces were found at the SiC/SiAlON and SiAlON/mullite interfaces. We estimated the approximate fracture toughness for the SiC/SiAlON and SiAlON/mullite interfaces and minimum limit of fracture toughness for the mullite/YbDS interface. Comparison between the fracture toughness by the experiment and calculated $\mathcal{G}_{init}^{th}$ and $\mathcal{G}_{prop}^{th}$ indicated that the fabricated T/EBC possesses sufficient mechanical reliability against interface cracks.

Of course, the present study considers only the stress state just after the fabrication process and does not assure the mechanical reliability in operation under thermal cycle with a high humidity, which is regarded as our future work.

Author Contributions: Conceptualization, E.K. and Y.U.; Fabrication of T/EBC, N.Y. and T.Y.; Measurement of material properties, N.Y., T.Y., H.K. and T.A.; FEM analysis, E.K. and A.K.; Interface fracture test, H.K.; Writing–original draft preparation for fabrication process and result, N.Y. and T.Y.; Writing–original draft preparation for FEM analysis and result, E.K. and A.K.; Writing–original draft preparation for interface fracture test procedure and result, H.K.; Writing–review and editing, S.K. and Y.U.; Supervision, H.K., T.A., S.K. and Y.U.; Project administration, S.K.; Funding acquisition, H.K., T.A., S.K. and Y.U.

Funding: This work was supported by Council for Science, Technology and Innovation (CSTI), Cabinet Office, Japan, through Cross-ministerial Strategic Innovation Promotion Program (SIP), "Structural Materials for Innovation" (Funding agency: JST, Japan Science and Technology Agency).

Acknowledgments: The authors would like to thank Kaori Sakata, Shoko Taniguchi and Takahumi Kawano for technical assistance.

Appendix A. Procedure for Calculating the Temperature-Dependent Material Properties

Appendix A.1. Young's Modulus, Poisson's Ratio and Shear Modulus

In this study, Young's moduli E at 303–1673 K of the SiC and SiAlON layers were considered to be identical to that of their bulk, E_{bulk}. E at the same temperature range of the mullite, YbDS and YbMS layers was estimated from the measured values of coating specimens deposited by the dual EB-PVD process with the same conditions described in Section 2.1, E_{film}. The outline of the estimation method of E of each layer is shown below;

Step 1: Derivation of an approximation formula of relation between temperature, T, and E_{bulk} of measured values or literature values.

Step 2: Estimation of E_{bulk} at the temperature range from 303 to 1673 K based on the approximation formula derived in Step 1.

Step 3: Measurement of E_{film} at 303 K of a film/substrate layered specimen.

Step 4: Estimation of temperature dependence of the E, $E(T)$, of the coating layers using the temperature dependence of E_{bulk} and the ratio of $E_{film}(303)$ to $E_{bulk}(303)$ shown in Equation (A1);

$$E(T) = E_{bulk}(T)\frac{E_{film}(303)}{E_{bulk}(303)}. \tag{A1}$$

Es of the SiC and SiAlON layers were estimated through Steps 1 and 2. The values of E of the mullite, YbDS and YbMS layers were estimated through Steps 1–4. Tables A1 and A2 summarize the estimation procedure for each layer at Step 1 and Steps 2–3, respectively. For the Yb-silicate gradient composition layer, E was obtained from Young's moduli of the YbMS and YbDS layers by the following Equation [17]:

$$E_{\beta:1-\beta} = (E_{YM})^{\beta}(E_{YD})^{1-\beta}, \ \beta = 0, 0.2, 0.5, 0.8, \text{ and } 1. \tag{A2}$$

Here, β and $1 - \beta$ denote volume content of YbMS and YbDS, respectively.

Table A1. Estimation procedure of E of SiC, SiAlON, mullite, YbDS and YbMS layers (Step 1).

Coating Layer	Step 1		
	Specimen (Bulk)	**Measurement Method**	**Temperature Range T [K]**
SC	Wafer [1]	JIS R1602	303–1573
SA	Sintered body [2]	Resonant method [5]	303–973
Mu	Sintered body [3]	JIS R1602	303–1273
YbDS	Sintered body [4]	JIS R1602	303–1273
YbMS	Reference [18]	Reference [18]	300–1600 [18]

[1] CVD-SiC wafer, ADMAP Inc.. [2] Raw powder: BSI3–001B, AG Materials Inc.. Sintering condition: 2023 K for two hours, applied pressure 40MPa, P_{N2} = 0.6MPa. [3] Raw powder: KM-101, KCM Corporation. Sintering condition: 1573 K for fifty hours in air → 2023 K for five hours in air. [4] Raw powder: self-build by ultrasonic spray pyrolysis. Sintering condition: 1673 K for two hours in air → Jet mill grinding → 1773 K for five hours in air. [5] Resonant method with cantilever bending and torsion mechanism. Dimensions of a specimen are 2 mm × 10 mm × 60 mm.

Table A2. Detail of estimation procedure of E of SiC, SiAlON, mullite, YbDS and YbMS layers (Steps 2 and 3).

Coating Layer	Step 2	Step 3	
	Approximation Formula	Specimen (Film/Substrate)	Measurement Method
SC	Linear	–	–
SA	Linear	–	–
Mu	$E = E_0 - FT \exp\left(\frac{-T_0}{T}\right)$ [1]	Mu/Mu	Nanoindentation test [2]
YbDS	$E = E_0 - FT \exp\left(\frac{-T_0}{T}\right)$ [1]	YbDS/YbDS	Nanoindentation test [2]
YbMS	$E = E_0 - FT \exp\left(\frac{-T_0}{T}\right)$ [1]	YbMS/YbDS	Nanoindentation test [2]

[1] Reference [19], Optimization of E_0, F, and T_0. [2] Oliver and Pharr method.

The values of ν of the SiC, SiAlON and YbDS layers were considered to be identical to those of their bulk. The outline of the estimation procedure of ν of each layer at Step 1 is shown in Table A3.

Step 1: Measurement of ν of the bulk, ν_{bulk} in some temperature range.

Step 2: Obtainment of ν in the temperature range from 303 to 1673 K by the assumption that ν_{bulk} outside the temperature range in Step 1 was identical to that measured at adjacent temperature.

Table A3. Estimation procedure of ν of SiC, SiAlON and YbDS layers (Step 1).

Coating Layer	Step 1		
	Specimen (Bulk)	Measuring Method	Temperature Range T [K]
SC	Description in Table A1	JIS R1602 and ASTM C848	Description in Table A1
SA	Description in Table A1	Description in Table A1	Description in Table A1
YbDS	Description in Table A1	JIS R1602 and ASTM C848	Description in Table A1

ν_{ij} of the anisotropic mullite layer in the temperature range from 303 to 1673 K was estimated using ν_{ij} of the bulk specimen described in Table A1 and E_{film} of the mullite layer derived from the procedure shown in Table A2. ν of the YbMS used in this estimation was literature value [9] and temperature-independent. For the Yb-silicate compositional gradient layers, ν was obtained from Young's moduli and bulk moduli, B, of YbMS and YbDS by the following equations:

$$\nu_{\beta:1-\beta} = \frac{1}{2}\left(1 - \frac{E_{\beta:1-\beta}}{3B_{\beta:1-\beta}}\right), \quad B_{\beta:1-\beta} = (B_{\text{YM}})^{\beta}(B_{\text{YD}})^{1-\beta}. \tag{A3}$$

The YbMS anisotropic homogeneous layer is characterized by anisotropic Young's modulus, E_i^{ani}, Poisson's ratio, ν_{ij}^{ani}, and shear modulus, G_{ij}^{ani} $(i, j = x, y, z; i \neq j)$. A method of calculations for these parameters will be discussed in Reference [14]. Shear modulus of mullite layer was estimated by Equation (A4).

$$G_{ij} = \frac{E_i E_j}{E_i + E_j + 2E_j \nu_{ij}}. \tag{A4}$$

Appendix A.2. CTE (Reference Temperature: 1673 K)

α' was calculated from CTE at the reference temperature of 303 K (α; evaluated as a function of temperature from experimental measurements) as follows:

$$\alpha'(T) = \frac{\alpha(T) \times (T - 303) - \alpha(1673) \times (1673 - 303)}{(T - 1673) \times (\alpha(1673) \times (1673 - 303) + 1)}. \tag{A5}$$

α of the SiC and SiAlON layers was considered to be identical to that of their bulk. The values of α of the sintered bodies and the wafer in the temperature range from 573 to 1673 K were measured based on JIS R1618. Then, α of the sintered bodies and the wafer at 323–1673 K was estimated by extrapolation of the relationship between α and T. α of the mullite, YbDS and YbMS layers at given temperature was estimated using measured values of film/substrate layered specimens deposited with the same conditions as in Section 2.1. The outline of the calculation procedure of α of each layer is described below;

Step 1: Derivation of relationship between strain, ε, and T for each film/substrate layered specimen (described in Table A2), based on digital image correlation method.

Step 2: Approximation of the ε–T relationship derived in Step 1 by Equation (A6).

$$\varepsilon = N_1(T - 303) + N_2(T - 303)^2. \tag{A6}$$

Step 3: Calculation of α in the temperature range from 323 to 1673 K using the approximation formula derived in Step 2.

The value of α of the Yb-silicate gradient composition layers was obtained by the law of mixture as follows [20]:

$$\alpha_{\beta:1-\beta} = \frac{\alpha_{YM}\beta B_{YM} + \alpha_{YD}(1-\beta)B_{YD}}{\beta B_{YM} + (1-\beta)B_{YD}}. \tag{A7}$$

In this study, α of the YbMS anisotropic homogeneous layer was regarded as identical to that of the YbMS dense layer.

Appendix A.3. Creep Properties

In this study, we assumed that the creep property of each layer follows the strain-hardening law expressed as

$$\dot{\varepsilon}^c = C\sigma^n, \tag{A8}$$

where $\dot{\varepsilon}^c$ denotes creep strain rate.

As described in Section 2.2.2, we assumed that creep was negligible for the SiC layer. Creep properties of the SiAlON, mullite, YbDS and YbMS layers were estimated by the method described below;

Step 1: Estimation of n at a given temperature and the average of n within the temperature range from the graph of the true stress and the strain rate derived from literature values and/or measured values. Noted that the average of n was used in the FEM analysis.

Step 2: Estimation of C at a given temperature using the average of n, true stress and strain rate derived in Step 1.

Step 3: Estimation of C in the temperature range from 1173 to 1673 K using C^* and Q/R in Equation (A9) based on C versus inverse of T plot derived in Step 2.

$$C = C^* \exp\left(-\frac{Q}{RT}\right). \tag{A9}$$

The estimation method of the creep properties of each layer at Step 1 is shown in Table A4. For the Yb-silicate gradient composition layers, C and n were obtained as follows:

$$C_{\beta:1-\beta} = (C_{YM})^\beta (C_{YD})^{1-\beta}, \tag{A10}$$

$$n_{\beta:1-\beta} = (n_{YM})^\beta (n_{YD})^{1-\beta}. \tag{A11}$$

The creep property of the YbMS anisotropic homogeneous layer was regarded as identical to that of the YbMS dense layer.

Table A4. Method of creep property estimation of the SiAlON, mullite, YbDS and YbMS layers (Step 1).

Coating Layer	Step 1		
	Specimen	Measurement Method	Temperature Range *T* [K]
SA	Reference [21]	Reference [21]	1723–1873
Mu	Reference [22]	Reference [22]	1638–1713
YbDS	Description in Table A1	Uniaxial compressive creep test [2]	1623–1723
YbMS	Sintered body [1]	Uniaxial compressive creep test [2]	1623–1723

[1] Raw powder: self-build by ultrasonic spray pyrolysis. Sintering condition: 1873 K for two hours in air → Jet mill grinding → 1823 K for 0.5 hours applied pressure 50 MPa in Ar, 1873 K for five hours in air. [2] Dimensions of a specimen are 3 mm × 2 mm × 2 mm.

Appendix B. Preliminary FEM Analyses: Determination of Objective Interfaces and Crack Lengths

As was explained in Introduction, an interface crack in the T/EBC is likely to occur at interfaces with strong thermal stress concentration due to the difference in the CTEs from the SiC/SiC substrate. Here we determined the objective interfaces and crack lengths from the stress distribution obtained by a preliminary FEM analysis. The thermal stress analysis was carried out in the way shown in Section 2.2.2 for the uncracked T/EBC model with the SiAlON and mullite layers both being 25 μm thick.

Figure A1 shows the distribution of σ_x near the interface edge in the constituent layers of the T/EBC after the cooling process. Thermal stress is found to be concentrated particularly at the SiC/SiAlON, SiAlON/mullite and mullite/YbDS interfaces. Thus, we focused on crack initiation and propagation at these objective interfaces in this study.

From the balance of the stress component in the *x* direction, we deduce $\sigma_x \sim 0$ at the interface edge and thus a dominating effect of σ_y on $\mathcal{G}^{th}_{init}$. Therefore, we determined the crack lengths examined to evaluate $\mathcal{G}^{th}_{init}$, based on the σ_y distribution in the vicinity of the interface edge. Figure A2 shows the σ_y distribution in the coating layers along the *y* axis at the positions r_x distant from the interface edge (r_x = 0, 5, 10, 60, 100, 300, 500 and 1000 μm). The highest level of σ_y was observed at the interface edge (r_x = 0 μm) while the σ_y distribution reduces to a negligible level inside the model. On the basis of the results, we examined the crack lengths of *a* = 2.5, 5.0 and 7.5 μm to evaluate $\mathcal{G}^{th}_{init}$. For evaluation of $\mathcal{G}^{th}_{prop}$, the crack length *a* was varied as *a* = 2.5, 5.0, 7.5, 10, 50, 100, 250, ..., 3500 μm in increments of 250 μm between 250 and 3500 μm.

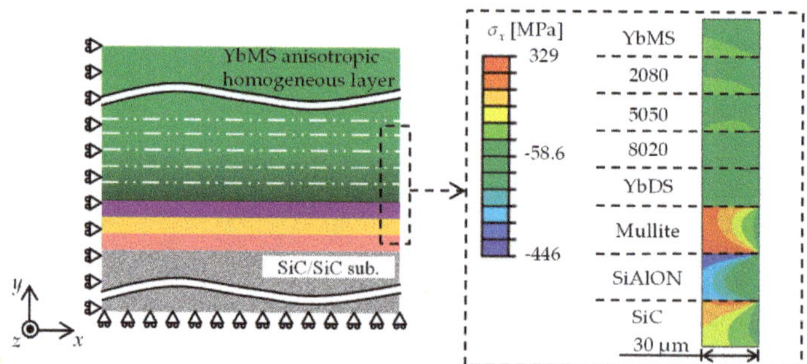

Figure A1. Thermal stress distribution near right interface edge of T/EBC deposited on SiC/SiC substrate without interface cracks.

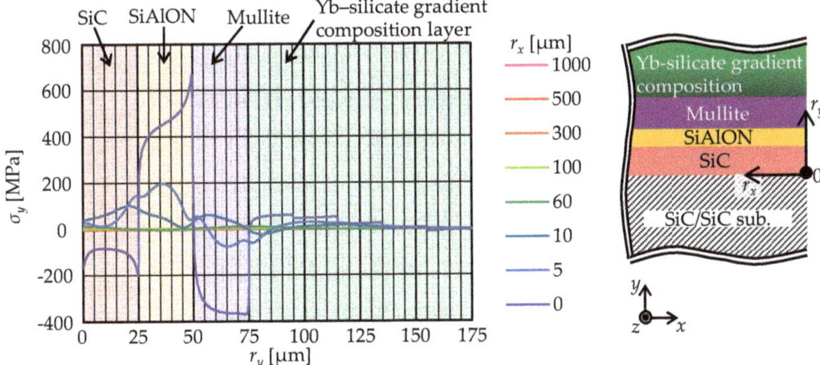

Figure A2. σ_y distributions in coating layers along the y axis at the positions r_x distant from right SiC/substrate interface edge in x direction.

Appendix C. Derivation Process of Energy Release Rate under Applying Mechanical Loading

Figure A3 shows the loads and moments applied to the specimen. P_b and M_b ($b = 1, 2$) are the load and moment from surrounding jigs, respectively, and P is the applied load by the pushing plate. Equilibria of forces and moments are given by the following equations, respectively [16,23,24]:

$$P - P_1 - P_2 = 0, \tag{A12}$$

and

$$P\left(\frac{h_{\text{coat}}}{2} + h_{\text{sub}} - \delta\right) - M_1 - M_2 = 0, \tag{A13}$$

where h_{sub} and h_{coat} are the thicknesses of the substrate and coating, respectively, and δ is the distance from the bottom of the substrate to the neutral axis of the specimen. When L is small compared to h_{sub}, large P_2 and M_2 can be generated, but in this case these are negligible. The specimen is a multi-layer composite of ten layers including the substrate with different Young's moduli, E_p, and thicknesses, h_p ($p = 1, 2, \cdots, 10$), so δ is given in a quite complicated form. Fortunately, in the loading condition for this test, the load applied to the coating is almost purely compressive parallel to the coating plane and the contribution of moment to the strain energy seems relatively small, thus the Young's modulus of the coating, E_{coat}, and the substrate, E_{sub}, may be simply approximated with the Voigt model in the rule of mixture, respectively,

$$E_{\text{sub}} = \frac{1}{h_{\text{sub}}} \sum E_s h_s, E_{\text{coat}} = \frac{1}{h_{\text{coat}}} \sum E_t h_t, s + t = 10, \tag{A14}$$

where s, t, h_{sub} and h_{coat} are varied according to the fractured interface. Thus δ can be obtained as the neutral axis position of a simple two-layer composite (single equivalent coating layer with E_{coat} and substrate with E_{sub}):

$$\delta = \Lambda h_{\text{coat}}, \Lambda = \frac{1 + 2\Sigma\eta + 2\Sigma\eta^2}{2\eta(1 + \Sigma\eta)}, \tag{A15}$$

where

$$\Sigma = \frac{E'_{\text{coat}}}{E'_{\text{sub}}}, \eta = \frac{h_{\text{coat}}}{h_{\text{sub}}}, E'_{\text{coat}} = \frac{E_{\text{coat}}}{1 - (\nu_{\text{coat}})^2}, E'_{\text{sub}} = \frac{E_{\text{sub}}}{1 - (\nu_{\text{sub}})^2}. \tag{A16}$$

ν_{coat} and ν_{sub} are Poisson's ratios of coating and substrate, respectively.

Energy release rate for the edge crack under an applied load of P is denoted by $\mathcal{G}$ and expressed as [24]

$$\mathcal{G} = \frac{1}{2w^2 E'_{\text{coat}}}\left(\frac{P^2}{h_{\text{coat}}}\right) + \frac{1}{2w^2 E'_{\text{sub}}}\left\{-\frac{(P_1)^2}{Dh_{\text{coat}}} - \frac{(M_1)^2}{I(h_{\text{coat}})^3} + \frac{(P_2)^2}{h_{\text{sub}}} + 12\frac{(M_2)^2}{(h_{\text{sub}})^3}\right\}. \qquad (A17)$$

Here, D is the dimensionless cross-sectional area and I is the moment of inertia of area written as

$$D = \frac{1}{\eta} + \Sigma,\ I = \Sigma\left\{\left(\Lambda - \frac{1}{\eta}\right)^2 - \left(\Lambda - \frac{1}{\eta}\right) + \frac{1}{3}\right\} + \frac{\Lambda}{\eta}\left(\Lambda - \frac{1}{\eta}\right) + \frac{1}{\eta^3}. \qquad (A18)$$

I is again calculated assuming that the material system is a two-layer composite with a coating and substrate. Substituting Equations (A12), (A13) and $P = P_{\text{init}}$ into (A17) and ignoring P_2 and M_2, Equation (4) is derived.

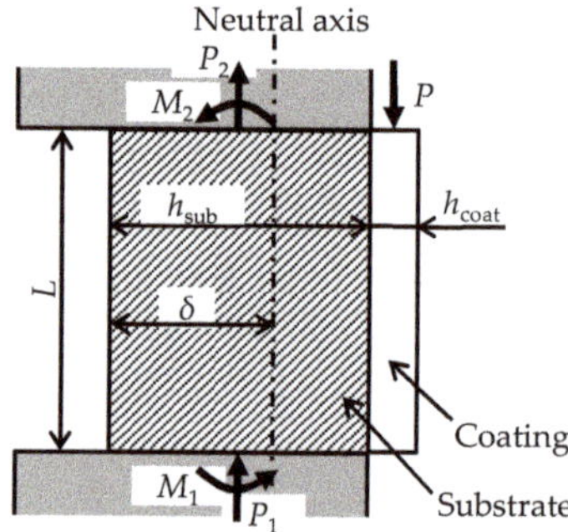

Figure A3. Loads and moments applied to specimen.

References

1. Pollock, T.M.; Tin, S. Nickel-based superalloys for advanced turbine engines: Chemistry, microstructure and properties. *J. Propul. Power* **2006**, *22*, 361–374. [CrossRef]
2. Spitsberg, I.; Steibel, J. Thermal and environmental barrier coatings for SiC/SiC CMCs in aircraft engine applications. *Int. J. Appl. Ceram. Technol.* **2004**, *1*, 291–301. [CrossRef]
3. Padture, N.P. Advanced structural ceramics in aerospace propulsion. *Nat. Mater.* **2016**, *15*, 804–809. [CrossRef] [PubMed]
4. Opila, E.J. Oxidation and volatilization of silica formers in water vapor. *J. Am. Ceram. Soc.* **2003**, *86*, 1238–1248. [CrossRef]
5. Robinson, R.C.; Smialek, J.L. SiC recession caused by SiO_2 Scale volatility under combustion conditions: I, experimental results and empirical model. *J. Am. Ceram. Soc.* **1999**, *82*, 1817–1825. [CrossRef]
6. Poerschke, D.L.; Van Sluytman, J.S.; Wong, K.B.; Levi, C.G. Thermochemical compatibility of ytterbia-(hafnia/silica) multilayers for environmental barrier coatings. *Acta Mater.* **2013**, *61*, 6743–6755. [CrossRef]
7. Richards, B.T.; Young, K.A.; de Francqueville, F.; Sehr, S.; Begley, M.R.; Wadley, H.N.G. Response of ytterbium disilicate-silicon environmental barrier coatings to thermal cycling in water vapor. *Acta Mater.* **2016**, *106*, 1–14. [CrossRef]
8. Klemm, H. Silicon nitride for high-temperature applications. *J. Am. Ceram. Soc.* **2010**, *93*, 1501–1522. [CrossRef]
9. Lu, M.-H.; Xiang, H.-M.; Feng, Z.-H.; Wang, X.-Y.; Zhou, Y.-C. Mechanical and thermal properties of Yb_2SiO_5: A promising material for T/EBCs applications. *J. Am. Ceram. Soc.* **2016**, *99*, 1404–1411. [CrossRef]
10. Kitaoka, S.; Matsudaira, T.; Yokoe, D.; Kato, T.; Takata, M. Oxygen permeation mechanism in polycrystalline mullite at high temperatures. *J. Am. Ceram. Soc.* **2017**, *100*, 3217–3226. [CrossRef]

11. Wada, M.; Matsudaira, T.; Kawashima, N.; Kitaoka, S.; Takata, M. Mass transfer in polycrystalline ytterbium disilicate under oxygen potential gradients at high temperatures. *Acta Mater.* **2017**, *135*, 372–381. [CrossRef]

12. Kitaoka, S.; Matsudaira, T.; Wada, M.; Yokoi, T.; Yamaguchi, N.; Kawashima, N.; Ogawa, T.; Yokoe, D.; Kato, T. Mass transfer in Yb silicates under oxygen potential gradients at high temperatures. *AMTC Lett.* **2019**, *6*, 150–151.

13. Kitaoka, S.; Matsudaira, T.; Kawashima, N.; Yokoe, D.; Takata, M. Structural stabilization of mullite films exposed to oxygen potential gradients at high temperatures. *Coatings* **2019**, *9*, 630. [CrossRef]

14. Kawai, E.; Kubo, A.; Umeno, Y. Homogenization method for calculating energy release rate in environmental barrier coatings with columnar layer for ceramics. manuscript in preparation.

15. The Society of Materials Science. *Strength and Fracture of Materials*, 4th ed.; The Society of Materials Science: Kyoto, Japan, 2011; p. 144. (In Japanese)

16. Aoki, Y.; Inoue, J.; Kagawa, Y.; Igashira, K. A simple method for measurement of shear delamination toughness in environmental barrier coatings. *Surf. Coat. Technol.* **2017**, *321*, 213–218. [CrossRef]

17. Kagawa, Y.; Hatta, H. *Tailoring Ceramic Composites*, 1st ed.; Agne Shofusha: Tokyo, Japan, 1990; pp. 96–97. (In Japanese)

18. Tian, Z.; Zheng, L.; Wang, J.; Wan, P.; Li, J.; Wang, J. Theoretical and experimental determination of the major thermo-mechanical properties of RE_2SiO_5 (RE = Tb, Dy, Ho, Er, Tm, Yb, Lu, and Y) for environmental and thermal barrier coating applications. *J. Eur. Ceram. Soc.* **2016**, *36*, 189–202. [CrossRef]

19. Wachtman, J.B., Jr.; Tefft, W.E.; Lam, D.G., Jr.; Apstein, C.S. Exponential temperature dependence of Young's modulus for several oxides. *Phys. Rev.* **1961**, *122*, 1754–1759. [CrossRef]

20. Turner, P.S. Thermal-expansion stresses in reinforced plastics. *J. Res. Natl. Bur. Stand.* **1946**, *37*, 239–250. [CrossRef]

21. Chihara, K.; Hiratsuka, D.; Tatami, J.; Wakai, F.; Komeya, K. High-temperature deformation of α-SiAlON nanoceramics without additives. *Scr. Mater.* **2007**, *56*, 871–874. [CrossRef]

22. Okamoto, Y.; Fukudome, H.; Hayashi, K.; Nishikawa, T. Creep deformation of polycrystalline mullite. *J. Eur. Ceram. Soc.* **1990**, *6*, 161–168. [CrossRef]

23. Suo, Z.; Hutchinson, J.W. Interface crack between two elastic layers. *Int. J. Fract.* **1990**, *43*, 1–18. [CrossRef]

24. Hutchinson, J.W.; Suo, Z. Mixed mode cracking in layered materials. In *Advances in Applied Mechanics*; Hutchinson, J.W., Wu, T.Y., Eds.; Elsevier: Amsterdam, The Netherlands, 1991; Volume 29, pp. 63–191.

MDPI
St. Alban-Anlage 66
4052 Basel
Switzerland
Tel. +41 61 683 77 34
Fax +41 61 302 89 18
www.mdpi.com

Coatings Editorial Office
E-mail: coatings@mdpi.com
www.mdpi.com/journal/coatings